General Chemistry
Guided Explorations

David M. Hanson
SUNY Stonybrook

Australia • Brazil • Canada • Mexico • Singapore • Spain • United Kingdom • United States

© 2008 Thomson Brooks/Cole, a part of The Thomson Corporation. Thomson, the Star logo, and Brooks/Cole are trademarks used herein under license.

ALL RIGHTS RESERVED. No part of this work covered by the copyright hereon may be reproduced or used in any form or by any means—graphic, electronic, or mechanical, including photocopying, recording, taping, Web distribution, information storage and retrieval systems, or in any other manner—without the written permission of the publisher.

Printed in the United States of America

1 2 3 4 5 6 7 11 10 09 08 07

Printer: ePAC

Cover Images: Background photo: Sean McKenzie/WWI/Peter Arnold, Inc.; inset photos (left to right): Clive Freeman/Biosym Technologies/Photo Researchers, Inc.; Courtesy of IBM Research, Almaden Research Center. Unauthorized uses prohibited.; Kenneth Eward/Photo Researchers, Inc.; Clive Freeman/Biosym Technologies/Photo Researchers, Inc.

ISBN-13: 978-0-495-11599-1
ISBN-10: 0-495-11599-1

Thomson Higher Education
10 Davis Drive
Belmont, CA 94002-3098
USA

For more information about our products, contact us at:
Thomson Learning Academic Resource Center
1-800-423-0563

For permission to use material from this text or product, submit a request online at
http://www.thomsonrights.com.
Any additional questions about permissions can be submitted by email to thomsonrights@thomson.com.

Table of Contents

	To the Student	v
01-1	The Nature of Matter	1
01-2	Scientists Love to Measure	5
02-1	The Nuclear Atom	11
02-2	The Mole and Molar Mass	13
03-1	Naming Compounds	19
03-2	Determining Molecular Formulas	23
04-1	Reaction Stoichiometry	27
04-2	Limiting Reactants	31
05-1	Types of Chemical Reactions	35
05-2	Solution Concentration and Stoichiometry	41
06-1	Energy	47
06-2	Enthalpy	51
07-1	Photoelectron Spectrum of Argon	55
07-2	Periodic Trends in Properties of Elements	63
08-1	Lewis Structures	69
08-2	Electronegativity and Bond Properties	73
09-1	VSEPR Model	77
09-2	Hybrid Atomic Orbitals	81
10-1	Gases and the Ideal Gas Law	87
10-2	Partial Pressures	91
11-1	Phases of Matter	95
11-2	Phase Diagrams	99
12-1	Fuels	103
12-2	Organic Functional Groups	107
13-1	Rates of Chemical Reactions	111
13-2	Reaction Mechanisms	119
14-1	Equilibrium Constant and Reaction Quotient	123
14-2	Calculating Equilibrium Concentrations	127
15-1	Solubility	131
15-2	Coligative Properties	137
16-1	Acid Ionization Constants	141
16-2	Calculations Involving Acid Ionization Constants	145

17-1	Buffers	149
17-2	Acid – Base Titrations	153
18-1	Entropy	157
18-2	Free Energy	163
19-1	Electrochemical Cells	165
19-2	Electrolytic Cells	169
20-1	Radioactivity	173
20-2	Rates of Nuclear Decay	175
21-1	Chemistry of Main Group Elements	179
21-2	Electronic Structure and Properties	181
22-1	Transition Metals and Coordination Compounds	183
22-2	Magnetism and Color in Coordination Compouds	187

To the Student

The activities in this book are designed to help you develop an understanding of some of the concepts that you will encounter in a General Chemistry course and to help you develop the ability to apply your understanding in answering questions and solving problems.

The activity design is based on findings from research in the cognitive sciences on how people learn. This research documents that people learn by connecting new knowledge to what they already know and by following a sequence of exploration, concept formation, and application.

To help you connect the new ideas to what you already know, the title of each activity includes a focus question and most activities begin with questions that ask for your opinion or ask you to make predictions.

The heart of each activity is the *Exploration*, where you are given a model to explore. The model is simply some representation of what you are to learn. It might be a diagram, a table of information, an illustrative problem, experimental data, or even some written text. Questions help you, guide your exploration of the model, and lead you to identifying and understanding the relevant concepts. The first few questions are directed. They point you to relevant information in the model. Subsequent questions ask you to put ideas together and come to conclusions, i.e. form and understand concepts.

You then practice applying your new knowledge in *exercises* and *problems*. The exercises are straightforward. They help you build confidence in using your new knowledge. Problems are more complex. Problems help you integrate your new knowledge with prior knowledge and apply it in new situations. You should work additional problems from your textbook as well.

At the end of each activity, a *Got It!* section helps you assess and document your success in learning the material. If you have trouble with this section, read the relevant material in your textbook, and go back and do the activity over again.

You will learn the most and have the most fun if you work on these activities with other students. Discussions among members of your learning team will produce different viewpoints regarding the concepts and their use in solving problems, will identify and correct misconceptions that you may have, and will strengthen and deepen your understanding of chemistry. Use your textbook to resolve disagreements, to find answers to questions that arise, and see examples of problem solutions.

01-1. The Nature of Matter: What is that substance?

Scientists classify matter as elements, compounds, pure substances, mixtures, and solutions. They also classify changes in matter as physical changes or chemical changes.

What do you think?

1. Apply the terms element, compound, pure substance, mixture, and solution to each of the following. More than one term may apply.
   (a) helium

   (b) table salt

   (c) blood

   (d) air

   (e) tea

2. Consistent with your classifications in item 1 above, describe what each of the following terms means to you.
   (a) element

   (b) compound

   (c) pure substance

   (d) mixture

   (e) solution

Information

The use of words in science is very precise. On the street, *oily dirt* may be a good answer to the question, "What is that substance?" In the science lab, however, *oily dirt* is not a substance, it is a mixture.

A *substance*, or more explicitly *pure substance*, is any pure matter that cannot be separated into components by physical methods like picking out the pieces, evaporation, filtration, distillation, or crystallization. The composition of a pure substance is always the same. Physical methods separate the components but do not change them. Chemical methods like combustion (burning) transform the substances into other substances.

Vodka is not a pure substance. Its composition, which is the amount of ethanol and water present, can vary, and it can be separated into ethanol and water by distillation. Ethanol, on the other hand, is a pure substance. Its composition always is the same, and it cannot be converted into any other substance by physical methods, but it can be converted into carbon dioxide and water by burning, which is a chemical process.

An *element* is a pure substance that can only be decomposed into other pure substances by nuclear reactions. An *element* cannot be decomposed into two or more other pure substances by either physical or chemical methods. All of the known elements are listed on the Periodic Table. Some examples of elements are hydrogen, helium, oxygen, nitrogen, silver, gold, and lead. When elements are combined, they can form mixtures or compounds, which are described below.

A *compound* is a pure substance formed from 2 or more elements. Glucose, a sugar, is a compound. It is formed from carbon, hydrogen, and oxygen.

A *mixture* consists of two or more pure substances. A mixture can be *homogeneous* or *heterogeneous*. A homogeneous mixture is uniform, the parts are not distinguishable, like sugar dissolved in water. Air, which is made up of oxygen, nitrogen, and small amounts of other gases, also is a homogeneous mixture. A heterogeneous mixture is not uniform, the parts are distinguishable, like salad dressing or a package of white and brown rice.

The term *mixture* generally applies to a mixture that is heterogeneous, and the term *solution* is used for a mixture that is homogeneous.

A *chemical change or process* involves the transformation of one or more pure substances into one or more different pure substances. A pure substance must be an element or a compound. Iron rusting is an example of a chemical change. Iron combines with oxygen to form rust.

A *physical change or process* involves changes in pure substances, mixtures, and solutions that do not transform the pure substances present into other pure substances. Water freezing or boiling and sugar dissolving in water are examples of physical changes.

Exploration

1. How do chemists use the term *substance* in a way that probably differs from the way *substance* is used in casual conversation?

2. Two elements can form a compound or a mixture. For example, hydrogen and oxygen can combine to form water; oxygen and nitrogen can combine to form air.
   (a) Which combination is a compound, and which is a mixture?

   (b) Is this mixture homogeneous or heterogeneous? Explain.

3. What are some similarities and differences between water vapor and air, both of which are made from two elements?

4. What is the difference between a physical process and a chemical process? Provide examples of each that are not given to you in this activity.

Application

1. Examine your previous classifications of each of the following and revise them as necessary in view of the new information that you have been given. Explain your rationale for each case.
   (a) helium

   (b) table salt

   (c) blood

   (d) air

   (e) tea

2. Identify each of the following as a physical or chemical change and explain your rationale.
   (a) the fender on a car rusts

   (b) water evaporates

   (c) sugar dissolves in water

   (d) water freezes at 0 °C

   (e) food is digested

Got It!

If you have mastered this material you should be able to

1. Explain the similarities and differences between pairs of the following terms: element, compound, substance, mixture, solution, physical change, and chemical change.

2. Provide examples for each of the terms in item 1 above.

3. Classify examples that you are given using the terms in item 1 above.

01-2. Scientists Love to Measure: Which athlete is heavier, taller, faster?

Scientists, as well as people in general, love to measure things. Extensive statistics are kept on the performance of athletes. Scientists use measurements to identify what is happening in experiments, and to verify or reject explanations and theories. Some of the basic quantities that you will encounter are listed in Table I.

What do you think?

1. Peyton and Eli Manning are two quarterbacks. Peyton plays for the Indianapolis Colts, and Eli plays for the New York Giants. Peyton weighs 105 kg, and Eli weighs 220 lbs. Who is heavier?

2. Candace Parker and Kia Vaughn are two basketball players. They starred in the 2007 Women's NCAA basketball tournament. Candace played for Tennessee, and Kia played for Rutgers. Candace is 76 in tall, and Kia is 1.93 m tall. Who is taller?

3. Jennifer Rodriguez and Apolo Ohno are two Olympic medalists and record holders in speed skating. Jennifer's time on the 500 m long track is 37.83 s. Apolo's time on the 500 m short track is 41,518 ms. Who is faster?

Table I. International System of Units Used in Measurements (SI Units)

Physical Quantity	Unit	Abbreviation
mass	kilogram	kg
length	meter	m
volume	liter	L
time	seconds	s
amount of substance	mole	mol
temperature	kelvin centigrade or celsius	K °C

Prefixes are added to units to deal with very large or very small numbers. The prefixes that you need to know are listed in Table II.

Table II. Prefixes Used with Units

Prefix	Abvtn	Meaning	Example with Scientific Notation
mega	M	10^6	1,000,000 Hz = 1.0×10^6 Hz = 1 MHz
kilo	k	10^3	1,000 g = 1.0×10^3 g = 1 kg
centi	c	10^{-2}	0.01 m = 1.0×10^{-2} m = 1 cm
milli	m	10^{-3}	0.001 L = 1.0×10^{-3} L = 1 mL
micro	μ	10^{-6}	0.000001 g = 1.0×10^{-6} g = 1μg
nano	n	10^{-9}	0.000000001 m = 1.0×10^{-9} m = 1 nm
pico	p	10^{-12}	0.000000000001 m = 1.0×10^{-12} m = 1 pm
femto	f	10^{-15}	0.000000000000001 s = 1.0×10^{15} s = 1 fs

Exploration - 1

1.1. What are the SI units for mass, length, and volume?

1.2. What are the meanings and abbreviations for the prefixes kilo, centi, milli, micro, and nano?

1.3. Which SI unit an d prefix would be most appropriate for measuring your body mass? Explain.

1.4. Which SI unit and prefix would be most appropriate to use with the diameter of a human hair? Explain.

1.5. What is the meaning and advantage of using scientific notation, e.g. 1.0×10^{-6}?

Information

Sometimes you need to convert from one unit to another. Engineers, health professionals, biologists, and other scientists find unit conversion to be a necessary part of their jobs. For example, the results of two measurements can only be compared directly if they have the same units.

For example, if you want to compare the price of gasoline in Europe (where gasoline is sold by the liter) and the United States (where gasoline is sold by the gallon), you need to convert liters to gallons, and in cooking it is helpful to be able to convert between cups and pints, pints and quarts, and teaspoons and tablespoons.

Also, when numerical values are calculated, the result must have the correct units. The units in the result are obtained by performing the arithmetic operations on the units as well as on the numbers. If the units obtained for the result are incorrect, then the value calculated also must be incorrect. Checking whether the units of the result are correct or not is a powerful method for validating a calculation. This validation is called *dimensional analysis* or *unit analysis*.

Unit conversion is accomplished by using equivalence statements to produce unit conversion factors. An *equivalence statement* is an equality that shows the relationship between two different units. A *conversion factor* is a ratio of units that equals 1. Since the conversion factor equals 1, it can multiply a quantity and change the units but not the actual physical magnitude of the quantity. Two conversion factors are obtained from an equivalence statement by dividing through by one side or the other.

For example, the equivalence statement for gallons and liters, Equation. 1, produces two conversion factors, Equations 2 and 3.

$$1 \text{ gallon} = 3.785 \text{ L} \qquad (1)$$

$$\frac{1 \text{ gal}}{3.785 \text{ L}} = \frac{3.785 \text{ L}}{3.785 \text{ L}} = 1 = 0.2642 \text{ gal/L} \qquad (2)$$

$$\frac{1 \text{ gal}}{1 \text{ gal}} = \frac{3.785 \text{ L}}{1 \text{ gal}} = 1 = 3.785 \text{ L/gal} \qquad (3)$$

The conversions factors in Equations (2) and (3) both equal 1 because the numerator and denominator of each represent the same thing; they just have different units.

Suppose we purchase 5.00 gal of gasoline in the United States and 18.93 L of gasoline in France. To compare the amount of gasoline purchased in each situation, we need to express the amounts in common units. So convert liters to gallons by using the conversion factor in Equation (2) to show that 18.93 L and 5.00 gal; are the same.

$$18.93 \text{ L } (0.2642 \text{ gal/L}) = 5.00 \text{ gal} \qquad (4)$$

Exploration - 2

2.1. How many unit conversion factors result from a single equivalence statement? Explain.

2.2. Which is the correct conversion factor to multiply by in order to convert liters to gallons, Eq. 1 or Eq. 2? Explain.

2.3. Why did the units in the result of the calculation shown by Equation (4) turn out to be gal as expected and not *L gal/L*?

2.4. (a) Multiply 15 gal by the conversion factor in Eq. 1 and do the arithmetic on the units as well as the numbers.

(b) Multiply 15 gal by the conversion factor in Eq. 2 and do the arithmetic on the units as well as on the numbers.

(c) Compare the results obtained in parts (a) and (b) and explain how doing the arithmetic on the units as well as the numbers identifies the correct conversion.

Application

1. The mass of a car is 1,365,000 g. Determine the mass in kg and in pounds. (1 kg = 2.205 lbs)
   (a) First estimate the answer without using a calculator, divide by 1000 to convert g to kg, then multiply by 2 to approximately convert kg to lbs. Write your estimate below. Making estimates is another way to validate your answers and inform you when you have made errors in the calculation.

   (b) Write the answer you obtained using a calculator.

2. A light pulse from a laser is 0.15 ns long. Determine the duration of this pulse in ps.
   (a) First identify whether a picosecond is longer or shorter than a nanosecond.

   (b) Based on your answer to (a) above, then identify whether the value in picoseconds will be larger or smaller than 0.15.

   (c) Now calculate your answer and validate it by comparing with your answer to (b).

3. Express the volume 1.0 L in terms of cubic meters, m^3. Note that 1.0 L = 1000 cm^3, and the length conversion factor is 1 m/100 cm.
   (a) Noting that the volume of a cube 10 cm on a side is 1000 cm^3 (10 cm x 10cm x 10 cm), why do you need to apply the length conversion factor three times to produce the volume conversion factor 1 m^3/10^6 cm^3?

(b) How many cubic meters are equivalent to 1.0L?

4. Estimate the number of atoms that could comprise a baseball, given that the volume of a baseball is 200 cm³ and the volume of an atom is 0.004 nm³. Do the arithmetic on the units as well as the numbers to validate that you have made the correct unit conversions.

5. You are the purchasing agent of a start-up biotechnology firm. If sucrose costs $11.80 per pound, and a bottle contains 5.00 kg, how much would you pay for a case of sucrose containing 12 bottles?

Got It!

1. Use the equivalence statement (1 kg = 2.205 lbs) to check your response to Question 1 in the *What do you think?* section. Write the mass of both football players in kilograms. Who is heavier?

2. Use the following equivalence statements to check your response to Question 2 in the *What do you think?* section. (1 m = 1.094 yd, 1 yd = 36 in) Write the height of both basketball players in meters. Who is taller?

3. Obtain the appropriate equivalence statement from Tables 1 and 2 to check your response to Question 3 in the *What do you think?* section. Write the time for each skater in seconds. Who is faster?

General Chemistry: Guided Explorations

02-1. The Nuclear Atom: What is the smallest particle of an element?

Atoms are the fundamental building blocks of all substances. Fig. 1 shows a representation of the atoms of three elements and gives the atomic symbols for them. Note that the nucleus and the surrounding electron cloud (e-) are not drawn to scale. The diameter of the electron cloud actually is 100,000 times larger than the diameter of the nucleus.

Exploration

1. What are the three particles that make up atoms?

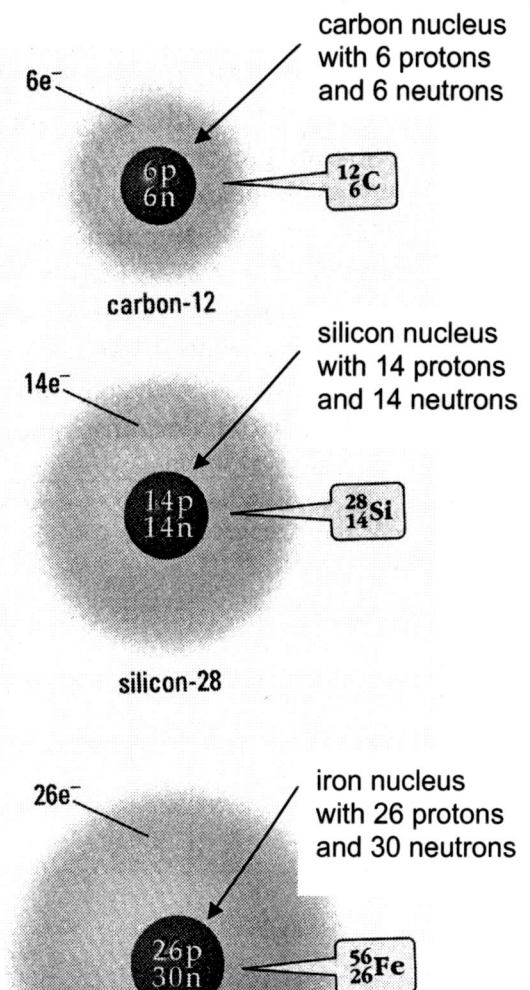

2. Where are each of the particles that make up atoms located in the atom?

3. What information is provided by the numbers in the atomic symbol, e.g. $^{56}_{26}Fe$?

4. What is the relationship between the number of protons and the number of electrons in any neutral (uncharged) atom?

Fig. 1 Atoms of carbon, silicon, and iron.

11

Information

All atoms of a given element have the same number of protons.

Atoms of the same element with different numbers of neutrons are called *isotopes*. Isotopes are distinguished by their different mass numbers, e.g. carbon-13 is an isotope of carbon-12.

Electrons can be removed from atoms producing an atomic ion that has a positive charge because it has more protons than electrons. A positively charged ion is called a *cation*. The charge of an ion is specified by a right superscript in the atomic symbol, e.g. Mg^{2+}.

Electrons also can be added to atoms producing a negatively charged ion. For example, O^{2-} has 2 more electrons than protons. A negatively charged ion is called an *anion*.

Got It!

1. Complete the entries in the Table I. The first row has been completed for you. In the table, Z = atomic number, which is the number of protons, A = mass number, which is the number of protons and neutrons, and N = number of neutrons. Use a Periodic Table to look up symbols, names, and atomic numbers where necessary.

Table I. Composition of Different Atoms

Name	Symbol	Z	A	N	Number of Electrons
Chlorine-37	$^{37}_{17}Cl$	17	37	20	17
Chloride	$^{37}_{17}Cl^-$				
	$^{35}_{17}Cl$				
Oxygen-16					
		6		6	
		6		7	
		11	23		10
Silver-107					46
Sulfur-32					18

General Chemistry: Guided Explorations

02-2. The Mole and Molar Mass: Can you count atoms and molecules by weighing them?

Atoms combine to form molecules in chemical reactions. Keeping track of the number of atoms of each element involved in a chemical reaction is very important. This knowledge enables chemists to determine molecular formulas and mechanisms of reactions. It also determines how much of each reactant to use in a chemical reaction and how much product to expect.

What do you think?

1. Is it possible to determine the number of objects from their mass?

2. If so, explain how the number of objects can be determined from their mass. If not, explain why not.

Exploration: Solving a problem and applying the solution to chemistry.

You purchased 10 pounds of pennies in small pail at an estate sale for $10.00. That seemed like a good deal, pennies for a dollar a pound. Rather than tediously counting all the pennies, you decide to determine how many you have from the mass. You find that a single penny weighs 2.509 g and that 1 kg = 2.205 lb.

1. How can you calculate the number of pennies in the pail from the information that is given?

2. Transform you answer to Question 1 above to a mathematical equation that shows how to calculate the number of pennies in the pail. Use the following symbols:
   N_p = the number of pennies,
   M_t = the total mass of all the pennies, and
   m_p = the mass of one penny.

3. How many pennies are in the pail that has 10 lbs of pennies?

4. (a) What is the value of the pennies in the pail?

   (b) Explain whether you got a good deal or not since you purchased the pail of pennies for $10.00.

Information

Masses of atoms can be determined by a technique called mass spectrometry. Since the mass of an individual atom is very small, a special unit is used. This unit is called an *amu* (atomic mass unit) with 1 amu = 1.6605×10^{-24} g. In dealing with naturally occurring samples, the average mass of all the isotopes is used because such samples contain all isotopes in their naturally occurring amounts.

5. Write an equation like the one you wrote for item 2 above that you can use to determine the number of atoms in a bar of platinum (Pt) given the mass of the bar and the average mass of a platinum atom.

6. How many atoms are there in a 1kg bar of platinum given that the average mass of a platinum atom is 195.08 amu?.

Information

In Question 6 above, you found that 1 kg of Pt contains 3.0871×10^{24} atoms. Dealing with such large numbers is not very convenient, so chemists invented a new unit for counting atoms. This unit is called a *mole*. This unit works just like the unit *dozen* for counting eggs. If you go to purchase a dozen eggs, you come back with 12 eggs. If you go to purchase a mole of platinum atoms, you come back with 6.0221×10^{23} platinum atoms. One mole of any substance is 6.0221×10^{23} particles of that substance. The number of particles in a mole is so important that it is given a special name and symbol. It is called *Avogadro's number* and has the symbol N_A.

The mass of one mole of a substance is called the *molar mass*. The molar mass always is given in grams not kilograms.

Application

1. If you purchase a dozen apples, how many apples do you get?

2. If you purchase a mole of apples, how many apples do you get?

3. If a dozen apples costs $3.00, how much does a mole of apples cost?

4. If a dozen apples weighs 2.0 kg, how much does a mole of apples weigh?

5. Show how to determine the molar mass (in grams) of Pt given that the average mass of a platinum atom is 195.08 amu.

6. Show how to determine the molar mass (in grams) of Cl_2, given that the average mass of a chlorine molecule is 70.90 amu.

7. From your answers to 1 and 2, identify the relationship between the molar mass in grams and the average mass in amu of a particle comprising a substance. Note that the molar masses of all the elements are listed on the Periodic Table.

8. In Exploration Question 6, you found that 1 kg of Pt contains 3.0871×10^{24} atoms.
(a) Show how you can use Avogadro's number with this information to determine the number of moles of Pt atoms in 1 kg of Pt.

(b) Show how you can use the molar mass of Pt to determine the number of moles of Pt atoms in 1 kg of Pt. You found the molar mass of Pt in Application Question 5 above.

Got It!

If you have mastered this material you should be able to convert between mass, moles, and number of particles. To demonstrate your mastery, complete the following statements using the terms *molar mass* and *Avogadro's number*, use dimensional or unit analysis to demonstrate that your answer is correct, and then apply your understanding in solving the problem at the end.

1. To convert grams to moles, divide by _____
   as shown by the following unit analysis.

2. To convert moles to number of particles multiply by_____
   as shown by the following unit analysis.

3. To convert number of particles to moles divide by_____
   as shown by the following unit analysis.

4. To convert moles to grams, multiply by _____
   as shown by the following unit analysis.

5. You are the purchasing agent for a pharmaceutical company that manufactures cisplatin, which is a potent chemotherapy drug. The molecular formula for cisplatin is $PtCl_2(NH_3)_2$. You need to order platinum, Pt, and chlorine, Cl_2, to use in the production of cisplatin. Since platinum is expensive and chlorine is toxic, you do not want to purchase more of either one than is needed. From the molecular formula you know that to make 1 molecule of cisplatin, you need 1 platinum atom, and 2 chlorine atoms.

(a) If you order 100 kg of platinum, how many moles of Pt will you receive?

(b) How many atoms of Pt will you receive?

(c) How many moles of chlorine, Cl_2, are needed to react with all the platinum?

(d) How many kg of chlorine are needed to go with the 100 kg of platinum?

(f) How many moles of cisplatin can be made from these amounts of platinum and chlorine?

General Chemistry: Guided Explorations

03-1. Naming Compounds: What's in a name?

Atoms combine to form molecules, for example, O + O → O_2 and C + O → CO. O_2 is a molecule of the element oxygen, and CO is a molecule of the compound carbon monoxide. A *compound* is a substance formed from two or more different elements.

In order to talk about compounds in a meaningful way, they need to have names that tell us something about their composition. The composition of a compound is represented by its molecular or chemical formula. In a molecular or chemical formula, the elements forming the compound are designated by the symbol for the element, and the number of atoms of that element in a molecule of the compound is given by a subscript. For example, CO_2, represents a molecule of the compound carbon dioxide that is composed of 1 carbon atom and 2 oxygen atoms.

What do you think?

Almost everyone knows that water is H_2O. Many waiters in restaurants write H_2O when you order water. Water is the common name for H_2O. Invent a name for water that reveals that it is composed of two hydrogen atoms and one oxygen atom.

Exploration – 1: Names of Binary Molecular Compounds

Table I. Names of Some Binary Molecular Compounds

Molecular Formula	Name
HF	Hydrogen fluoride
HCl	Hydrogen chloride
HBr	Hydrogen bromide
H_2S	Hydrogen sulfide
CO	Carbon monoxide
CO_2	Carbon dioxide
CCl_4	Carbon tetrachloride
N_2O	Dinitrogen oxide
N_2O_5	Dinitrogen pentoxide
PBr_3	Phosphorous tribromide
PBr_5	Phosphorous pentabromide
SF_6	Sulfur hexafluoride

1.1. As evident from Table I, how many different elements are there in a binary compound?

1.2. Which element is named first, the one that is less electronegative or the one that is more electronegative? The more electronegative elements are to the right on the Periodic Table. *Electronegativity* refers to the attraction an atom has for electrons.

1.3. What ending (suffix) is applied to the root of the more electronegative element in the second part of the name?

1.4. What prefixes are used to indicate, when necessary, that the number of atoms is 1, 2, 3, 4, 5, or 6?

Information

Some elements always combine with others in certain ratios so it is not necessary to specify explicitly the number atoms of each element in the compound. For example, hydrogen always combines with the halogens (Group VIIA) in a 1:1 ratio, and with the chalcogenides (Group VIA) in a 2:1 ratio. So water need not be called *dihydrogen oxide*, just *hydrogen oxide* is adequate.

Not all compounds are molecular. Some are ionic. In an ionic compound, one or more electrons are transferred from one atom to another to form ions, see Figure 1. This transfer occurs because one atom is much more electronegative than the other.

The atom that lost the electrons has a positive charge and is called a *cation*. The atom that gained electrons has negative

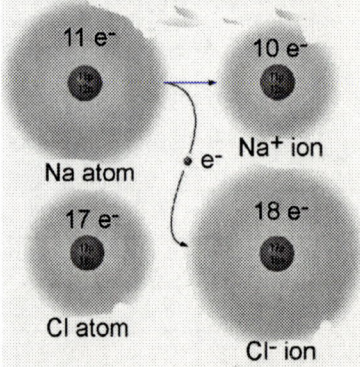

Fig. 1. Formation of an ionic compound NaCl.

20

charge as a result and is called an *anion*. The two ions are held together by the electrostatic attraction between the positive and negative charges.

Cations are formed from metals that are not very electronegative on the left side of the Periodic Table. Anions are formed from the electronegative nonmetals on the right side of the Periodic Table.

Exploration – 2: Names of Binary Ionic Compounds

Table II. Names of Some Binary Ionic Compounds

Chemical Formula	Name	Cation Charge	Anion Charge
NaCl	Sodium chloride		
KBr	Potassium bromide		
MgCl$_2$	Magnesium chloride		
BaI$_2$	Barium iodide		
CaO	Calcium oxide		
CuCl	Copper(I) chloride		
CuCl$_2$	Copper(II) chloride		
CrCl$_3$	Chromium(III) chloride		
Al$_2$O$_3$	Aluminum oxide		
Fe$_2$O$_3$	Iron(III) oxide		
FeO	Iron(II) oxide		

2.1. According to Table II, which element is named first in an ionic compound, the metal or the nonmetal?

2.2. What ending (suffix) is applied to the root of the nonmetal in the second part of the name?

2.3. When a metal ion can form more than one kind of cation, how is the charge on the ion indicated?

4. What are some similarities and differences in the names of binary molecular compounds and binary ionic compounds?

Information

Group IA and Group IIA elements always give up 1 and 2 electrons, respectively.

When forming ionic compounds, oxygen and other Group VIA elements always take 2 electrons, and the halogens in Group VIIA always take 1 electron.

Some transition metals can lose varying numbers of electrons to form cations with different charges. The possibilities you should know are Cr (+2 and +3), Fe (+2 and +3), Co (+2 and +3), and Cu (+1 and +2). In a compound's name, the charge on the metal is indicated by the Roman numerals in parentheses following the name of the metal.

Got It!

If you have mastered this material, you should be able to complete the following.

1. Write the charges on the cations and anions in the appropriate columns of Table II.

2. Write the formulas for the following compounds.
   (a) sulfur dioxide

   (b) silicon tetrachloride

   (c) chromium(III) oxide

3. Write the names of the following compounds.
   (a) $CoCl_2$

   (b) $CoCl_3$

   (c) K_2S

   (d) SO_3

4. Explain why it is not necessary in Table II to name
   (a) $MgCl_2$ magnesium dichloride.

   (b) $CuCl_2$ copper(II) dichloride

03-2. Determining Molecular Formulas: Why can different molecules have the same percent composition by mass?

Exploration

Your challenge is to complete Table I below and identify what the pairs of molecules listed in the table have in common. You should be able to do this right after you have written the molecular formulas, but also calculate the percent composition by mass as well. Write the percent composition using 3 significant figures. Use the first row to check that your procedure is correct since this row for acetylene has been done for you.

. To complete this task, you need to be able to convert structural formulas into molecular formulas. You also need to be able to determine the molar mass of compounds, and the percent by mass of each element in 1 mole of the compound

Table I. Compounds with Something in Common

Name	Structural Formula	Molecular Formula	Percent by Mass Carbon	Hydrogen
Pair 1				
Acetylene	H−C≡C−H	C_2H_2	92.2%	7.76%
Benzene	(hexagonal ring with 6 H)			
Pair 2				
Ethene	H₂C=CH₂			
Cyclohexane	(hexagonal ring)			

23

1. What are the common features of the four compounds in Table I that you can identify from the table?

2. How are the molecular formulas of these compounds similar and how are they different?

3. If a chemical analysis provides you with the mass percent composition of an unknown compound, could you determine the molecular formula of the compound from this information? Explain why or why not based on the information in Table I.

4. Why do the two compounds in Pair 1 and the two compounds in Pair 2 have the same mass percent composition?

5. The formula that expresses the correct ratio of carbon atoms to hydrogen atoms for both compounds in Pair 1 is CH. What is the corresponding formula that expresses the correct ratio of carbon atoms to hydrogen atoms for both compounds in Pair 2?

General Chemistry: Guided Explorations

Information

The formulas in Question 5 above, CH and CH$_2$, are called empirical formulas. An *empirical formula* for a compound gives the symbols for the elements in the compound, and the subscripts provide the ratio of the atoms of each element using the smallest possible whole numbers (integers). The molecular formula is some multiple of the empirical formula. The empirical formula can be determined from the mass percent composition, but additional information, for example, the molar mass, is needed to determine the molecular formula from the empirical formula.

Application

Phosphorous forms two compounds with oxygen. One contains 56.34% phosphorous and 43.66% oxygen. Use this information to determine its empirical formula.

To obtain the empirical formula, we need the ratio of O atoms to P atoms in the compound. The mass percent composition tells us that if we had 100 g of the compound, 56.34 g would be P and 43.66 g would be O. We can use this information to get the ratio that we need.

1. How many moles of P atoms are there in 56.34 g of P? Show that the answer is 1.819 mol P.

2. How many moles of O atoms are there in 43.66 g of O? Show that the answer is 2.729 mol O.

3. What is the smallest whole number ratio of O atoms to P atoms? From Items 1 and 2 we see that the ratio is
$$\frac{\text{O atoms}}{\text{P atoms}} = \frac{2.729}{1.819}.$$
We need to convert these numbers into whole numbers without changing the value of the ratio, so divide both the numerator and the denominator by the smallest,
$$\frac{\text{O atoms}}{\text{P atoms}} = \frac{2.729/1.819}{1.819/1.819} = \frac{1.500}{1.000},$$

and then multiply both the numerator and denominator by 2.

$$\frac{\text{O atoms}}{\text{P atoms}} = \frac{1.500}{1.000} = \frac{2 \times 1.500}{2 \times 1.000} = \frac{3}{2}$$

4. What is the empirical formula for this compound?

5. Show that the molar mass associated with the empirical formula is 109.95 g/mol.

6. The molar mass of this compound was determined to be 219.9 g/mol. How many multiples of the empirical formula does it take to produce a mass of 219.9 g/mol?

7. What is the molecular formula of this compound?

Got It!

Another oxide of phosphorous contains 43.64% phosphorous and 56.36% oxygen. It has a molar mass of 283.89 g/mol. Determine its empirical and molecular formulas.

04-1. Reaction Stoichiometry: How much do I need to get what I want?

Determining the amounts of reactants and products in a chemical reaction is called *reaction stoichiometry*. Stoichiometry is a word that comes from the Greek and means *measuring the pieces*. Balanced reaction equations are used to relate the amounts of products produced by various amounts of reactants.

You have a stoichiometry problem whenever you are asked, *How much is produced? How much is used? How much is needed? What is the reaction yield? What is the limiting reactant?* To answer such questions, you need to determine the quantity of one reactant or product from the quantity of another reactant or product.

There are three keys to stoichiometry problems. You need to be able to (1) balance the reaction equation, (2) recognize that the amounts in moles of products and reactants are in the same proportions as the stoichiometric coefficients in the balanced equation, and (3) make various conversions. *What are these conversions?* you ask. They are summarized in Table I. Don't memorize this table, rather use unit analysis to figure out the conversion.

Table I. Conversions Found in Stoichiometry Problems

Quantities	Conversion Factor	Unit Analysis
grams to moles	Divide by molar mass	g / (g /mol) = mol
moles to grams	Multiply by molar mass	mol x g/mol = g
gas volume to moles	Multiply by P/(RT).	L x mol/L = mol
moles to gas volume	Multiply by RT/P.	mol x L/mol = L
liquid density to mass	Multiply by the volume.	(g/mL) x mL = g
solution volume to moles of solute	Multiply by the molar concentration.	L x (mol/L) = mol

A specific strategy, like following the steps below, can be very helpful in solving stoichiometry problems.

Step 1: Write the balanced reaction equation if it is not given to you.

Step 2: Write the amounts you are given and identify what you need to find immediately below the reaction equation. Include units!

Step 3: Convert quantities to moles and write the values below the reaction equation as well.

Step 4: Set up a mole ratio involving the quantity of the compound that you need to find and the quantity of the compound that you is given. Set up another ratio involving the stoichiometric coefficients of these two compounds in the balanced reaction equation. These two ratios must be equal because the moles of reactants and products must be in proportion to the stoichiometric coefficients.

Step 5: Solve for what you need to find in the ratio, and convert moles to the units requested.

Example

The reaction below is used to produce nitrogen gas to inflate air bags in automobiles. How many grams of sodium azide are needed to inflate a 50 L airbag to pressure of 2.0 atm at 300 K.

Step 1	$10\ NaN_3(s) + 2\ KNO_3(s) \rightarrow 16\ N_2(g) + K_2O(s) + 5\ Na_2O(s)$
Step 2	? g $\qquad\qquad\qquad\qquad\qquad$ $N_2(g)$ = 50L at 2.0 atm and 300 K
Step 3	$\qquad\qquad\qquad\qquad\qquad$ n = PV/(RT) = 4.1 mol of $N_2(g)$, see below
Step 4	$\dfrac{10\ \text{mol sodium azide}}{16\ \text{mol nitrogen}} = \dfrac{x}{4.1\ \text{mol nitrogen}}$ $\rightarrow$ x = 2.6 mol sodium azide
Step 5	2.6 mol × 65 g/mol = 170 g sodium azide needed

Calculation of moles of nitrogen gas needed.
PV = nRT so n = PV/(RT)
n = (2.0 atm)(50 L) / [(0.08206 L atm mol^{-1} K^{-1})(300 K)]
n = 4.1 mol

Exploration

1. Examine the five step strategy for solving stoichiometry problems that is followed in the above example.

    (a) Why is writing the balanced reaction equation an important part of solving stoichiometry problems?

    (b) What purpose is served by Steps 2 and 3?

    (c) In Step 4, why must the mole ratio of the amounts equal the ratio of the stoichiometric coefficients?

2. Are there any steps in the strategy that you feel are not useful? Explain.

3. Can you improve the strategy by changing the steps or changing the order of the steps? Explain.

4. What are 5 insights about solving stoichiometry problems have you gained by reading the introductory material and examining the strategy and its use in the example?

Application

1. When 0.5 mol of ammonium nitrate decomposes, how many moles of water are produced?

   ___NH_4NO_3(s) → ___N_2(g) + ___H_2O(g) + ___O_2(g)

2. How many moles of ammonium nitrate must decompose to produce 96 g of oxygen?

3. How many grams of a Freon, CCl_2F_2, are produced by reacting 231 g of carbon tetrachloride, CCl_4, with hydrofluoric acid, HF?

Got It!

1. Cisplatin, $PtCl_2(NH_3)_2$, is a drug used in the treatment of cancer. It can be synthesized by the reaction of K_2PtCl_4 with ammonia. How many liters of ammonia gas is needed at STP to produce 1.0 kg of cisplatin?

2. Nitroglycerin is unstable and decomposes with explosive violence because it releases much energy and forms a large amount of gaseous products. How many moles of gas will be produced by 227 g of nitroglycerin exploding?

   4 $C_3H_5(NO_3)_3$(l) → 12 CO_2(g) + 10 H_2O9g) + 6 N_2(g) + O_2(g)

04-2. Limiting Reactants

Reactants are not always present in the exact amounts indicated by a balanced reaction equation. Usually in laboratory reactions or industrial processes, one reactant is more expensive or less readily available than others. The cheaper or more available reactants are used in excess to ensure that the more expensive or rarer substance is completely converted to product.

The limiting reactant is the reactant that is completely converted to product. Once the limiting reactant has been used up, no more product can form. Given information about the amounts of reactants present, you need to be able to determine which one is the limiting reactant. The moles of product formed are determined by the number of moles of the limiting reactant present at the beginning of the reaction.

As demonstrated below, an analogy for the limiting reactant can be made with the ingredients of a cheese sandwich.

What do you think?

In making cheese sandwiches or other food dishes, there often is a limiting ingredient. Suppose you have 40 bread slices, 25 cheese slices, and 30 lettuce leaves, how many cheese sandwiches could you make? Each sandwich must have 2 slices or bread, 1 slice of cheese, and 1 leaf of lettuce. This recipe can be written like a reaction equation.

2 bread slices + 1 cheese slice + 1 lettuce leaf → 1 sandwich

Exploration

1. Which ingredients were left over after you made the cheese sandwiches?

2. Which ingredient was the limiting ingredient?

3. What procedure did you use to identify the limiting ingredient?

4. Try your procedure and see if it works with molecules. Molecules are just like cheese sandwiches. Suppose you have 4 nitrogen molecules and 6 hydrogen molecules, how many ammonia molecules can you make?

    $N_2 + 3H_2 \rightarrow 2NH_3$

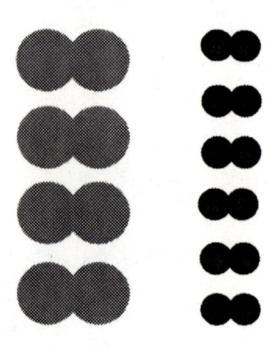

    Fig. 1. 4 N_2 molecules and 6 H_2 molecules

5. How can you use the stoichiometric coefficients in the balanced reaction equation help you identify the limiting reactant?

Information

One way to identify the limiting ingredient or limiting reactant is to see how much product can be made by each. For example, 40 bread slices can make 20 sandwiches, 25 cheese slices make 25 sandwiches, and 30 lettuce leaves make 30 sandwiches. The bread slices make the fewest number of sandwiches so they are the limiting ingredient.

Similarly 8 ammonia molecules can be made from 4 nitrogen molecules, and 4 ammonia molecules can be made from 6 hydrogen molecules, so in Question 5, hydrogen is the limiting reactant because it produces the fewest ammonia molecules.

Another way to identify the limiting reactant is to compare the ratio of the stoichiometric coefficients in the reaction equation to the ratio of the molecules or moles of reactants available. For example,

$$\text{stoichiometric coefficents} = \frac{1 \ N_2}{3 \ H_2}$$

$$\text{amounts available} = \frac{4 \text{ mol } N_2}{6 \text{ mol } H_2}$$

Since reactants must react in the same ratio as the stoichiometric coefficients, the fact that 4/6 = 2/3 is larger than 1/3 means that there is too much nitrogen, and hydrogen is the limiting reactant.

Application

1. Sketch a diagram in the empty space under Exploration Question 4 to show the ammonia molecules and the left over nitrogen molecules. Use circles shaded differently to represent the N and H atoms.

2. A rocket fuel, hydrazine H_2H_4, is produced by the reaction of chlorine, Cl_2, with sodium hydroxide, NaOH, and ammonia, NH_3. How many moles of hydrazine can be produced using excess sodium hydroxide, 15 mol chlorine, and 15 mol ammonia.

$$2NaOH + Cl_2 + 2NH_3 \rightarrow N_2H_4 + 2NaCl + 2H_2O$$

Got It!

Dialuminum hexachloride, Al$_2$Cl$_6$, is used in many industrial processes. It is made by treating scrap aluminum metal (Al) with chlorine gas (Cl$_2$).

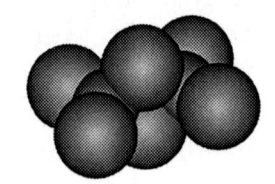

1. Write the balanced reaction equation for the production of dialuminum hexachloride.

2. What is the limiting reactant if 710 g of chlorine gas reacts with 270 g of aluminum?

3. How much dialuminum hexachloride is produced by this reaction?

4. How much of each reactant is left over when the reaction is complete?

5. If only 750 g of dialuminum hexachloride was produced, what was the percent yield of the reaction? Note: per cent yield = (actually amount produced / maximum amount that could be produced) times 100.

General Chemistry: Guided Explorations

05-1. Types of Chemical Reactions: What happens in a chemical reaction?

What do you think?

Describe what you think happens in a chemical reaction, and identify different types of chemical reactions that you may have encountered.

Information

Chemical compounds can react with each other and form new compounds. You need to visualize or represent these reactions in three ways: macroscopic, nanoscopic, and symbolic. The macroscopic representation is what you see with your eyes. The nanoscopic representation is what the atoms and molecules are doing, and the symbolic representation is a chemical reaction equation. Most likely you can find pictures and diagrams in your textbook that show what chemical reactions look like at the macroscopic and nanoscopic levels. This activity is directed at helping you understand reaction equations and identify some different types of chemical reactions.

Exploration

1. Some different types of chemical reaction are listed Table I. In order to make sense of the information in this table, you need to understand the symbols and words that are used. Use your textbook to provide an explanation for each of the following.

    (a) The symbols (s), (g), (l), and (aq).

    (b) ionic compound.

(c) dissolve.

(d) solution.

(e) acid and base

(f) hydronium ion and hydroxide ion.

Table I. Types of Reactions

Reaction Types / Characteristics	Examples
Dissociation an ionic compound dissolves in water to produce ions	$Ca(NO_3)_2(s) \rightarrow Ca^{2+}(aq) + 2NO_3^-(aq)$ *calcium nitrate dissociates*
Precipitation a solid forms when solutions are mixed	$AgNO_3(aq) + KCl(aq) \rightarrow AgCl(s) + KNO_3(aq)$ *silver nitrate reacts with potassium chloride*
Acid Ionization an acid reacts with water to produce hydronium ions	$HCl(g) + H_2O(l) \rightarrow H_3O^+(aq) + Cl^-(aq)$ *hydrochloric acid ionizes*
Base Ionization a base reacts with water to produce hydroxide ions	$NH_3(g) + H_2O(l) \rightarrow NH_4^+(aq) + OH^-(aq)$ *ammonia ionizes*
Base Dissociation a base dissolves in water to produce hydroxide ions	$NaOH(s) \rightarrow Na^+(aq) + OH^-(aq)$ *sodium hydroxide dissociates*
Acid – Base Neutralization an acid and a base react to produce water and a salt	$HCl(aq) + NaOH(aq) \rightarrow H_2O(l) + NaCl(aq)$ *hydrochloric acid neutralizes sodium hydroxide*
Oxidation – Reduction electrons are transferred from one reactant to another	$2Ag^+(aq) + Cu(s) \rightarrow 2Ag(s) + Cu^{2+}(aq)$ *electrons are transferred from copper atoms (oxidation) to silver ions (reduction)*

2. What are the characteristics of the reactions in Table I that you can use to identify the reaction? Identify the nature of the reaction and the general types of substances that react and are produced for each of the following reactions. You need to remember these characteristics, and writing them without looking at Table I will help you remember.

(a) dissociation

(b) precipitation

(c) acid ionization

(d) base ionization

(e) base dissociation

(f) acid-base neutralization

(g) oxidation – reduction

3. Which 3 types of reactions in Table I involve the transfer of a hydrogen ion, H⁺, from one reactant to another?

4. Explain how the oxidation reduction equation in Table I can be balanced when the charge on the silver ion is +1 and the charge on the copper ion is +2.

5. In the precipitation example in Table I, the reactants are ionic compounds so they dissociate in water. Consequently KCl(aq) means K⁺(aq) + Cl⁻(aq) and AgNO₃(aq) means Ag⁺(aq) + NO₃⁻(aq). Write the reaction equation using only those ions the produce the precipitate AgCl(s).

Information

Notice that all the reaction equations in Table I are balanced. *Balanced* means that the number of atoms of each element and the number of electrons are the same on both the reactant side and the product side of the equation. Reaction equations must be balanced because electrons and atoms are not created or destroyed in a chemical reaction.

The overall reaction equation is written for the precipitation example in Table I. All the compounds in this example are ionic so in aqueous solution they all dissociate except for silver chloride, AgCl, which is not soluble in water. Consequently KCl(aq) means K⁺(aq) + Cl⁻(aq) and AgNO₃(aq) means Ag⁺(aq) + NO₃⁻(aq). The precipitation reaction involves only the silver ions and the chloride ions. In the reaction equation that is written below, the potassium ions and the nitrate ions are not included because they are not involved in the precipitation reaction. These ions that are not involved are called *spectator ions*. A reaction equation that includes only those ions that are involved is called the *net ionic equation*. The net ionic equation, which should be your answer to Question 5 above, for the precipitation of silver chloride is

$$Ag^+(aq) + Cl^-(aq) \rightarrow AgCl(s).$$

As with precipitation reactions, net ionic equations can be written for acid-base neutralization reactions. For the example given in Table I, hydrochloric acid ionizes and sodium hydroxide dissociates in aqueous solution, so the net ionic equation is

$$H_3O^+(aq) + OH^-(aq) \rightarrow 2H_2O(l),$$

and the sodium and chloride ions are spectator ions.

Application

Complete and balance the following reactions, and identify the reaction type.

1. $NaSO_4(s) \rightarrow \quad Na^+(aq) \quad + \quad SO_4^{2-}(aq)$

2. $CO_3^{2-}(aq) \quad + \quad H_2O(l) \quad \rightarrow \quad H_2CO_3(aq) \quad + \quad OH^-(aq)$

3. $Fe^{3+}(aq) \quad + \quad OH^-(aq) \quad \rightarrow \quad Fe(OH)_3(s)$

4. $H_3PO_4(aq) \quad + \quad NaOH(s) \quad \rightarrow \quad Na_3PO_4(aq) \quad +$ _____

5. $HOCl(aq) \quad + \quad H_2O(l) \quad \rightarrow$ _____ $+ \quad ClO^-(aq)$

6. $Ca(s) \quad + \quad O_2(g) \quad \rightarrow \quad CaO(s)$

7. $Sr(OH)_2(aq) \quad \rightarrow \quad Sr^{2+}(aq) \quad +$ _____

Got It!

1. Lead(II) nitrate and hydrochloric acid react to produce a lead chloride precipitate.

   (a) Write the dissociation equation for lead(II) nitrate.

   (b) Write the acid ionization equation for hydrochloric acid.

   (c) Write the net ionic equation for the precipitation of lead(II) chloride.

2. Write the dissociation equation for sodium carbonate, Na_2CO_3, and the net ionic equation for the reaction of the carbonate ion, which is a base, with water.

3. A zinc metal rod, Zn, immersed in a solution of copper sulfate, $CuSO_4(aq)$, will dissolve producing Zn^{2+} ions in solution and copper metal will form on the rod. What type of reaction is this? Write the net ionic equation for it.

4. Write the equations that describe what happens (a) when nitric acid is added to water, (b) when sodium hydroxide is added to water, and (c) when the two solutions are mixed to neutralize each other.

General Chemistry: Guided Explorations

05-2. Solutions: What are they, what do they look like, and how are they prepared?

What do you think?

Identify three solutions that you have encountered that involve water as the solvent.

Exploration 1: Electrolytes and Nonelectrolytes

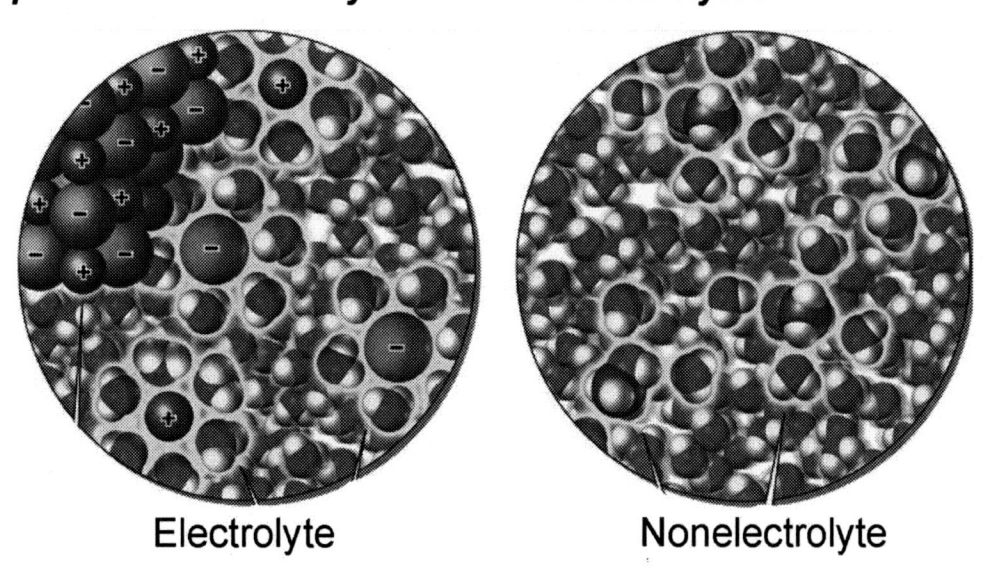

Electrolyte   Nonelectrolyte

1.1. The above diagram represents sodium chloride, NaCl, which is an ionic compound, and methanol, CH_3OH, which is a covalent compound, dissolving in water. Look carefully and identify the sodium and chloride ions, methanol molecules, and water molecules in this diagram. Which compound is an electrolyte and which one is the nonelectrolyte? Identify one similarity and one difference between these two types of solutions.

41

1.2. Which part of the water molecule is closest to the sodium ions in solution? Explain why.

1.3. Which part of the water molecule is closest to the chloride ions in solution? Explain why.

Information

Many of the materials that you encounter are solutions. Some examples of solutions are air, blood, beverages, and metal alloys like brass. A *solution* is a homogeneous mixture of two or more substances. The *solvent* is the component of the solution that is present in the largest amount. An *aqueous solution* is one that has water as the solvent. The *solute* is the substance, which is present in a smaller amount, dissolved in the solvent. There may be more than one solute in a solution. The *concentration* of a solution specifies the amount of solute present in some amount of solution or solvent.

Solutions can be in a solid, a liquid, or a gas. Examples of these are steel, gasoline, and air, respectively. The solute dissolved in a liquid solution initially can be a solid like sodium chloride, a liquid like ethanol, or a gas like oxygen. These dissolve in water to make salt water, an alcoholic beverage, or oxygenated water that supports fish life.

Substances that dissolve in water can be classified according to their ability to conduct electricity when in solution. These substances are *strong electrolytes* if they conduct electricity well, *weak electrolytes* if they conduct electricity poorly, and *nonelectrolytes* if they do not conduct electricity.

Many ionic compounds are strong electrolytes. When an ionic compound like sodium chloride, NaCl, dissolves in water, the cation and anion separate, and the motion of these charged particles through the solution serves to conduct electricity.

Covalent compounds like methanol or sugar are nonelectrolytes because when they dissolve in water they do not separate into charged particles. All the atoms remain covalently bonded to each other.

Some compounds are in between these two extremes of ionizing completely or not at all. Only a fraction of the molecules ionize, and only a few ions are produced that serve to conduct electricity for these weak electrolytes.

Exploration 2: Preparing Solutions of Known Concentration

2.1. The picture to the right shows that 7.410 g of calcium hydroxide, Ca(OH)$_2$ are being added to a volumetric flask to make 250.0 mL of of solution. How many moles of calcium hydroxide is this?

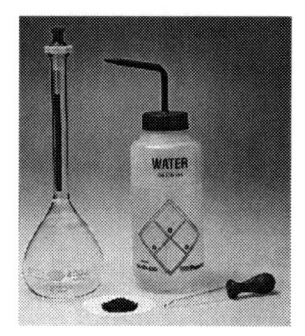

2.2. If 7.410 g of calcium hydroxide are used to make 250.0 mL of solution,
(a) what is the concentration of this solution in moles/liter of solution?

(b) How many moles of calcium ion would there be in one liter of a calcium hydroxide solution with this concentration?

(c) How many moles of hydroxide ion would there be one liter of a calcium hydroxide solution with this concentration?

Information

Solution concentration is commonly expressed as moles of solute in one liter of solution. This concentration is called the *molarity* of the solution and is represented by the letter M or by square brackets [ ] around a molecular formula. M stands for moles per liter. For example, [NaOH] = 1.5 M, and 1.5 M is read as 1.5 molar, which means that there are 1.5 moles per liter of solution. The molarity of a solution is calculated using the following formula, where n_{sol} = moles of solute and V_{soln} = volume of solution.

$$M = n_{solute} / V_{solution} \qquad (1)$$

2.3. The picture to the right shows that a 100.0 mL volumetric flask has been filled with 0.100 M potassium dichromate solution, $K_2Cr_2O_7$. This solution is to be transferred to the large 1.00 L volumetric flask, which then will be filled with water to the 1.00 L mark, producing 1.00 L of potassium dichromate solution.

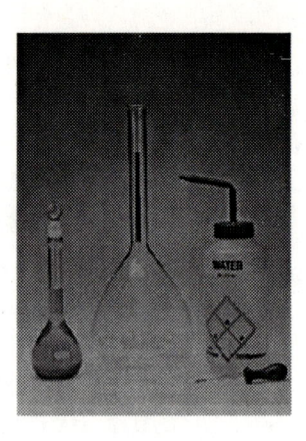

(a) How many moles of potassium dichromate are in 100.0 ml of 0.100 M solution?

(b) How many moles of potassium dichromate will end up in the 1.00 L volumetric flask?

(c) What will the molar concentration of the potassium dichromate solution be in the 1.00 L flask after water has been added to the 1.00 mark?

(d) What will the concentration of potassium ions be?

Information

In preparing solutions by dilution of a more concentrated solution, the moles of solute present in the concentrated solution is n, where n = M_cV_c as seen by rearranging Equation (1). M_c is the molarity of the concentrated solution, and V_c is the volume of the concentrated solution. These moles of solute end up in the diluted solution, so n also must equal M_dV_d, where M_d is the molarity of the diluted solution, and V_d is the volume of the diluted solution. This conclusion can be represented by Equation (2). Given any three of the quantities in Equation (2), the remaining one can be calculated.

$$M_cV_c = M_dV_d \qquad (2)$$

General Chemistry: Guided Explorations

Exploration 3: Relating Amounts of Reactants and Products in Solution

How many mL of a 3.0 M hydrochloric acid solution are needed to exactly neutralize 250.0 mL of 6.0 M sodium hydroxide solution?

3.1. Write the balanced reaction equation for this acid base reaction.

3.2. How many moles of sodium hydroxide are there in 250.0 mL of the 6.0 M solution?

3.3. How many moles of hydrochloric acid are needed to exactly neutralize the amount of sodium hydroxide you reported in Question 3.2?

3.4. What volume of the 3.0 M hydrochloric acid solution will contain the number of moles of hydrochloric acid that you reported in Question 3.3?

Information

You can solve solution stoichiometry problems like the one in Exploration 3 in the same way that you solve other stoichiometry problems. The following five step strategy is recommended.
(1) Write the balanced reaction equation.
(2) Write the amounts you are given, and what you need to find below each compound in the reaction equation.
(3) Convert the amounts to moles. In solution problems, you generally need to multiply volume in liters times molar concentration.
(4) Use the ratio of stoichiometric coefficients to find what you need to find.
(5) Convert moles to the units requested. In solution problems, if volume is requested, you divide the number of moles by the molar concentration.

Got It!

1. Identify one of the following as a strong electrolyte, one as a weak electrolyte, and one as a nonelectrolyte.

   propanol, $CH_3CH_2CH_2OH$

   acetic acid, CH_3COOH

   nitric acid, HNO_3

2. Identify the ions present in a solution of 0.12 M $BaCl_2$ and their concentrations.

3. Calculate the molarity of 250.0 mL of solution made from 0.25 moles of ammonium nitrate.

4. Calculate the concentration of a solution produced by diluting 10.0 mL of 12.5 M hydrochloric acid to 500.0 mL.

5. Determine how much 12.0 M HCl you need to use to produce 250.0 mL of a 3.00 M solution.

6. What volume of 2.06 M $KMnO_4$ solution contains 322 g of solute?

7. What mass in grams of $Ba(OH)_2$ is required to react completely with 25.0 mL of 0.125 M HNO_3?

8. Vitamin C has the formula C_6H8O_6, and can be oxidized by bromine, Br_2 to give HBr and $C_6H_6O_6$. It took 27.85 mL of 0.102 M Br_2 solution to react completely with a 1.00 g tablet of Vitamin C.
   (a) What is the mass in grams of vitamin C in the tablet?
   (b) What percent of the tablet is vitamin C?

06-1 Energy: How do you know how much you have gained or lost?

Richard Feynman, a Nobel laureate in physics, said energy is a mysterious thing; "...we have no knowledge of what energy is." [R.P.Feynman, R.B. Leighton, and M. Sands, *The Feynman Lectures on Physics* (Addison-Wesley, 1963) pp. 4-1 to 4-2]

Einstein recognized that mass, m, and energy, E, are related by the speed of light, c. He proposed that

$$E = mc^2. \qquad (1)$$

This equation can be transformed to $m = E/c^2$ and interpreted to mean that mass is just another form of energy. From this perspective, energy is one of the fundamental components that make up the universe, along with elementary particles like quarks.

The concept of energy is useful because it is a numerical quantity that is conserved. No matter what happens, no matter what changes occur, the numerical quantity called energy does not change. This idea, which is called the *Law of Conservation of Energy*, leads to the dictionary definition of energy.

A dictionary definition says that *energy* is a measure of the capacity to do work. Since energy is conserved, the concept of energy enables us to keep track of the capacity to do work.

Work is defined in science as a force applied over some distance. Since there are different kinds of forces, there appear to be different forms of energy (e.g. gravitational, electrical, chemical, nuclear, radiation, thermal, acoustic, and mechanical).

These different forms of energy can be classified into two types: kinetic energy and potential energy. *Kinetic energy* is the energy associated with the motion of an object. *Potential energy* is the energy associated with the position of an object.

The bottom line is that no one really understands what energy is, but we have ways of calculating energy, and whenever we add all the energy together, we always get the same number. This result means that energy is conserved. *Conservation of energy* means that energy is neither created nor destroyed, but it can be converted from one form to another.

Energy is measured in units of joules, J. *Joule* rhymes with *rule* and was the name of an English physicist, who established that the energy of a system can be changed by heating and doing work. This idea, which now is known as the *First Law of Thermodynamics* is expressed as

$$\Delta E = q + w \qquad (2)$$

where ΔE is the change in energy of the system, q is the energy transferred into the system, and w is the work that is done on the system.

Exploration: Measuring the Amount of Energy

Fig. 1. Transferring energy to water by heating.

As shown Fig. 1, an iron rod is heated over a Bunsen burner. It then is plunged into water. The temperature of the water rises from 18.1°C to 27.4°C, at which point the iron rod and water are at the same temperature. This exploration will determine the amount of energy transferred from the iron rod to the water.

1. What is the change in temperature of the water caused by heating the water with the hot iron bar?

2. If the beaker had half as much water in it, do you think that the temperature change would be the same, twice as large, half as large, or some other amount? Explain.

3. If the beaker had ethylene glycol in it rather than water, do you think the temperature change would be the same or different? Explain.

Information

The specific heat capacity of a substance is the amount of energy required to increase the temperature of 1 g of that substance by 1°C. Specific heat capacities for some substances are given in Table I.

Table I. Specific Heat Capacities

Symbol or Formula	Substance	Specific Heat Capacity J / g·°C
Fe	Iron	0.451
Al	Aluminum	0.902
$H_2O(l)$	Liquid water	4.184
$HOCH_2CH_2OH$	Ethylene glycol	2.42

4. How can the specific heat capacity be combined with other information to calculate the amount of energy transferred to a substance by heating?

5. Transform your answer to Question 4 into a mathematical equation using the following symbols: ΔT = change in temperature, c = specific heat capacity, m = mass, q = energy transferred.

Application

1. If there is 500. mL of water in the beaker in Fig. 1, how much energy was transferred from the iron rod to the water?

2. If this same amount of energy were transferred to 250. mL of water, what would the change in temperature of the water be?

49

3. If this same amount of energy were transferred to 500. mL of ethylene glycol, what would the change in temperature be?

4. Determine the final temperature of your 110 g aluminum cup when heated by 1.1 kJ of energy. The cup initially was at 23°C.

5. An acid-base neutralization reaction causes the temperature of a solution and beaker to rise from 23°C to 73°C. The volume of the solution is 400. mL. Determine the amount of energy released by this reaction. Assume that the specific heat capacity of the solution is the same as water and that the beaker has a heat capacity of 300. J/°C.

Got It!

1. Identify which will be hotter if you heat equal masses of iron and aluminum with 1 kJ of energy starting at room temperature. Explain.

2. You are holding an Al rod in one hand and a Fe rod in the other. The rods have the same mass and initial temperature (273 K). Assuming the rate of heat transfer is the same for both, which rod will reach your body temperature first? Explain.

General Chemistry: Guided Explorations

06-2. Enthalpy: How much heat is produced by a chemical reaction under different conditions?

Exploration

Fig. 1 shows a diagram of a calorimeter that can be operated in either a constant volume mode (A) or in a constant pressure mode (B). In the constant volume mode, a bar is inserted to hold the piston in place. In the constant pressure mode, a pressure is applied that resists a change in volume.

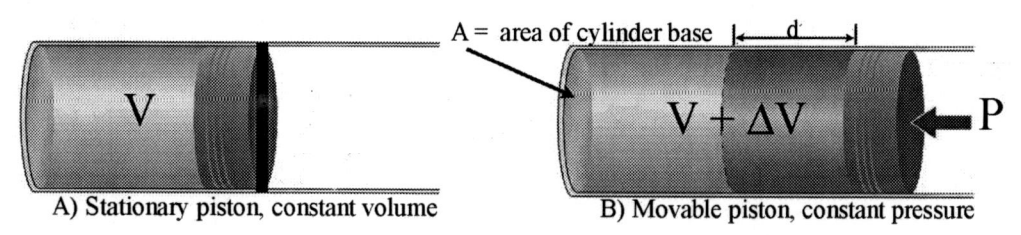

Fig. 1. Combustion in a calorimeter operated at (A) constant volume and (B) constant pressure. The calorimeter heat capacity is 25.370 kJ/°C.

Two experiments are conducted with liquid octane: one in the constant volume mode, and one in the constant pressure mode. Octane, or its equivalent, is the principal ingredient in gasoline. In each experiment, 11.42 g of liquid octane (C_8H_{18}, 114.2 g/mol) is burned completely in excess oxygen to produce $CO_2(g)$ and $H_2O(g)$. The change in temperature of the calorimeter is measured in each case. The results of these experiments are given in Table I.

Table I. Results of the Calorimetry Experiments with Octane.

Experiment	Sample mass	Initial temp	Final temp	ΔT
#1 constant V	11.42 g	21.560 °C	41.607 °C	
#2 constant P	11.42 g	23.690 °C	43.690 °C	

1. Write the balanced reaction equation for the combustion of one mole of liquid octane to produce carbon dioxide and water, both in the gas phase.

2. Calculate the temperature change of the calorimeter for experiments #1 and #2 and enter the values you obtain in the last column of Table I, include both values and units.

51

3. Show how to use the heat capacity of the calorimeter, which is given in the figure caption, to calculate the energy transferred to the calorimeter in experiment #1 and experiment #2. Enter your results in Table II, include both values and units (5 sig figs). Also enter the moles of octane consumed in each experiment.

Table II. Summary of the Calorimetry Experiments with Octane.

Experiment	Octane consumed (moles)	Energy transferred, q
#1 constant V		
#2 constant P		

4. In what ways were experiment #1 and experiment #2 different?

5. Why would a difference that you cited in answer to Question 4 lead to a difference in the amount of energy transferred to the calorimeter that you reported in Table II?

6. In view of the First Law of Thermodynamics, $\Delta E = q + w$, which experiment, #1 or #2, gives you directly the change in energy of the reacting system (oxygen and octane) from the data in Table II? Explain.

7. See the *Information* section below to learn how to calculate the amount of work done in experiment #2 as the system expanded against the external pressure. How much work was done by this expansion?

General Chemistry: Guided Explorations

8. In terms of the First Law of Thermodynamics, $\Delta E = q + w$,
   (a) did the energy of the reacting system increase or decrease in experiments #1 and #2?

   (b) is the sign of q positive or negative in these experiments? Go back and add the sign to the q values in Table II.

   (c) is the sign of w that you calculated in Question 7 positive or negative? Go back and make sure you have the correct sign to your answer to Question 7.

9. Show that the work w_p plus the energy transferred to the calorimeter q_p in experiment #2 equals the change in energy q_v determined in experiment #1. Explain why these two quantities are equal. Make sure you are using the correct signs with q_v, q_p, and w_p.

Information

If energy, q, is transferred into a system or work, w, is done on a system, then the energy of the system increases. The change in energy of the system, ΔE, is given by Equation 1, which is called the *First Law of Thermodynamics*.

$$\Delta E = q + w \qquad (1)$$

If energy is transferred out of the system or work is done by the system, then the energy of the system decreases. Equation 1 also describes this situation, except then q and w are negative (q < 0 and w < 0) because the energy, E, decreases.

Work is done when a force moves some object through some distance. If nothing moves, then no work is done, w = 0, and

$$\Delta E = q_v. \qquad (2)$$

The subscript in q_v indicates that energy is transferred with the volume constant and no work being done.

Physical processes and chemical reactions in plants, animals, laboratories, and our environment generally do not occur in closed containers with a constant volume. Instead they occur in contact with the atmosphere at constant pressure. If there is a change in volume, the atmosphere has to be pushed back. Work is done in pushing back the atmosphere, and this work is called expansion work.

Expansion work is determined by the change in volume and the pressure, P, resisting the expansion. The change in volume is given by $\Delta V = V_{final} - V_{initial}$. When $V_{final} > V_{initial}$, work has been done by the system, so the energy of the system must decrease, and w in equation (1) is negative.

$$w_p = - P \Delta V \qquad (2)$$

The subscript in w_p indicates that only expansion work is done.

The volume change usually is significant only when gases are produced or consumed in a chemical reaction. The ideal gas equation then can be used to relate $P\Delta V$ in Equation (2) to the change in then number of moles of gas at constant pressure as shown by Equations (3) through (5).

$$PV = nRT \qquad (3)$$

$$\text{so} - P \Delta V = - \Delta n \, RT \qquad (4)$$

$$\text{and} \; w_p = - \Delta n \, RT \qquad (5)$$

Since generally processes occur at constant pressure and not constant volume, the energy transferred between a system and its surroundings at constant pressure is a very important quantity. When the only work done is expansion work, the energy transferred at constant pressure is attributed to a change in enthalpy of the system, ΔH.

$$q_p = \Delta E - w_p = \Delta H \qquad (6)$$

Experiment #1 gave you the change in energy, ΔE, for the combustion of octane, and experiment #2 gave you the change in enthalpy, ΔH. They are different because of the expansion work that you calculated in Question 7.

Got It!

Show how to calculate the change in enthalpy from the data in Table II for one mole of liquid octane reacting with oxygen to produce gaseous carbon dioxide and water.

07-1. Photoelectron Spectrum of Argon: How do we learn about the electronic structure of atoms?

Before beginning this activity, you need to read your textbook to learn about electromagnetic radiation, the photoelectric effect, the Bohr model of the hydrogen atom, and the quantum numbers n, l, and m_l that are used to identify the electron atomic orbitals and energy levels in atoms. The relevant information is summarized below.

The lowest energy state of an atom or molecule is called the *ground state*. The absorption of electromagnetic radiation by matter can excite atoms and molecules to higher energy states and even ionize electrons.

Electromagnetic radiation has both wave properties and particle properties. The wave properties include a frequency, ν, wavelength, λ, and speed, c. These quantities are related by the following equation.

$$\nu \lambda = c \qquad (1)$$

The particle properties include momentum and the fact that energy is delivered in packets called photons. The energy of a photon is given by

$$E_{photon} = h\nu \qquad (2)$$

Visible light is one region of the electromagnetic spectrum. Visible light extends from a wavelength of 400 nm to 700 nm. X-rays form another region of the electromagnetic spectrum. X-rays have very short wavelengths and high photon energies.

In photoelectron spectroscopy, photons are used to ionize electrons from atoms, molecules, and solids. This phenomenon was first discovered in the early 1900s and is called the photoelectric effect.

A photoelectron spectrum is a plot or graph of the kinetic energy of the emitted electrons and the number of electrons emitted per second at each kinetic energy.

Photoelectron spectra of argon are shown in Figures 1 and 2. Such spectra can be obtained by ionizing argon atoms with x-ray photons that have an energy of 3500 eV.

Electron volts (eV) is the energy unit that is preferred by scientists who use this technique because the joule unit is too large and an electron volt is a more appropriate size (1eV = 1.602×10^{-19} J).

The key equation used to analyze photoelectron spectra is obtained by considering that energy must be conserved. The energy of the photon is used to overcome the binding energy of the electron in the atom, molecule, or solid, and any excess energy goes into the kinetic energy of the emitted electron.

$$E_{photon} = E_{binding} + E_{kinetic} \tag{3}$$

Electron atomic orbitals are identified by quantum numbers. The principal quantum number, n, has integer values (n = 1, 2, 3, 4...). As n increases, the energy of the electron increases, and the electron on average is further away from the nucleus and less tightly bound to it. For each value of n, the azimuthal quantum number, l, has integer values that range from 0 up to a maximum of n − 1.

Are You Prepared to Explore?

1. What is the wavelength of an x-ray photon that has an energy of 3500 eV?

2. What happens in the photoelectric effect?

3. What quantities are plotted on the x and y axes of a photoelectron spectrum?

4. What happens to the energy of an electron as the value of the principal quantum number increases?

Exploration

Photoelectron Spectrum of Argon
hv = 3500 eV

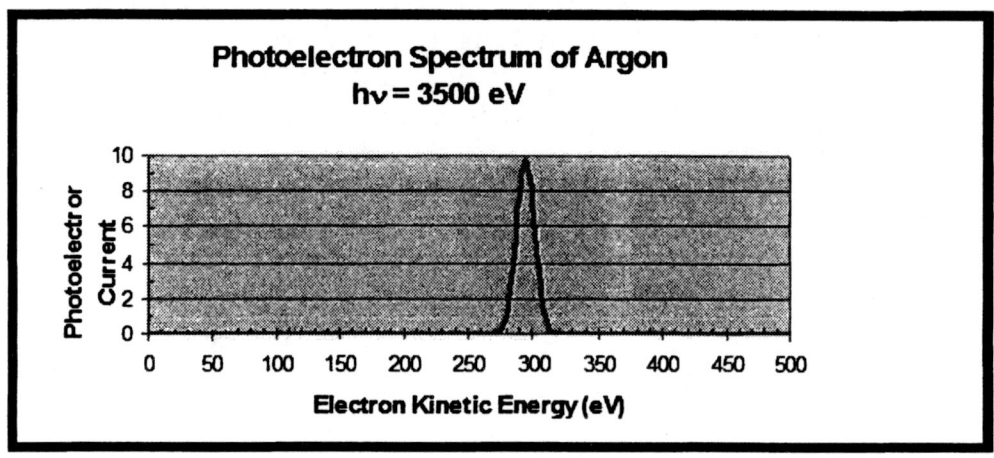

Fig. 1 Photoelectron spectrum of argon from 0 to 500 eV.

Photoelectron Spectrum of Argon
hv = 3500 eV

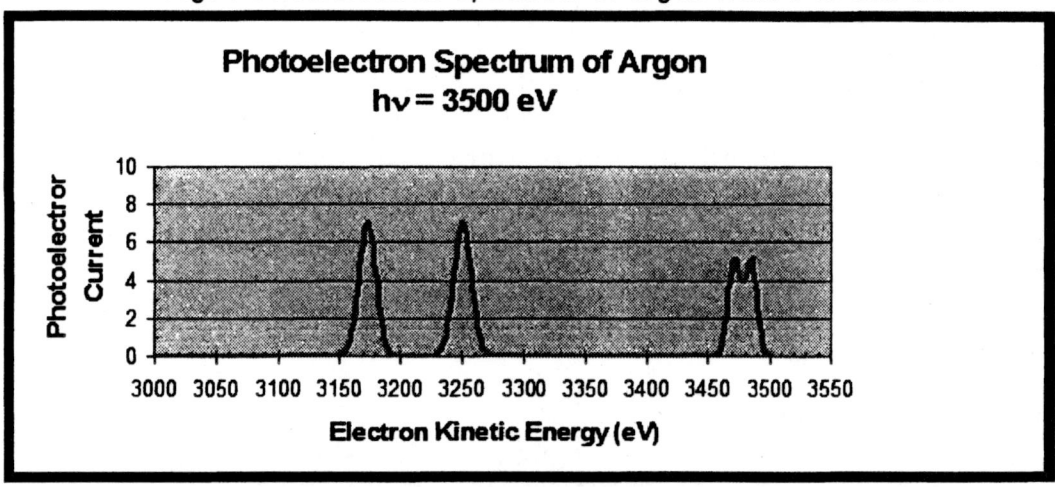

Fig. 2 Photoelectron spectrum of argon from 3000 to 3550 eV.

1. Peaks in the above photoelectron spectra occur at electron kinetic energies of 295, 3175, 3250, 3470, and 3485 eV. Label each peak in Figures 1 and 2 with the appropriate value of the electron kinetic energy: 295, 3175, 3250, 3470, and 3485 eV.

2. Complete Table I, below.

(a) Rearranging Equation (3) shows that the electron binding energy can be obtained by subtracting the electron kinetic energy from the photon energy. Using the data in Figures 1 and 2, determine the electron binding energies and write the values in the Table I

$$E_{binding} = E_{photon} - E_{kinetic} \qquad (4)$$

(b) Check that the sum $E_{kinetic} + E_{binding}$ in each column of Table 1 adds up to 3500 eV. Explain why this sum must be 3500 eV.

(c) The different peaks in the photoelectron spectrum are produced by ionizing electrons with different binding energies. The most tightly bound electrons, those with the largest binding energy, are in orbitals that have the smallest values of n and l. Write the values of n and l in Table I to assign the peaks in the photoelectron spectrum. The possible values are (n,l) = (1,0), (2,0), (2,1), (3,0), and (3,1).

Table I. Argon Photoelectron Data

$E_{kinetic}$(eV)	3485	3270	3250	3175	295
$E_{binding}$(eV)					
n value					
l value					

3. Using the data in Table I, complete the energy level diagram below to show the binding energy of electrons in argon. Label each energy level with the value of the binding energy and the values for the quantum numbers n and l. The lowest energy level is done for you.

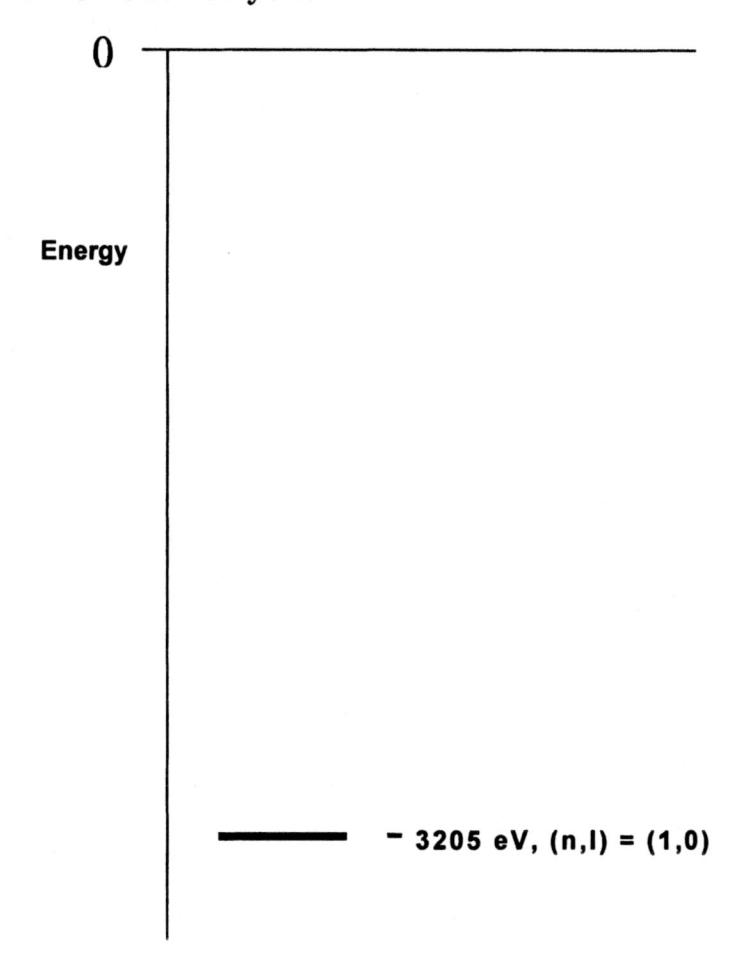

```
0 ─────────────

    Energy





                    ──── ⁻3205 eV, (n,l) = (1,0)
```

Information

The photoelectron spectrum and energy level structure can be understood in terms of a shell model for the atom. Electrons in the n=1 shell are closest to the nucleus and are strongly bound by the +18 charge of the argon nucleus. Electrons in the n=3 shell are furthest from the nucleus and are only weakly bound. Electrons in the n=2 shell are intermediate between these two. There are 3 shells with n = 1, 2, and 3 but 5 peaks in the photoelectron spectrum. The n=2 and n=3 shells are split into subshells with each subshell identified by a different value for l.

Application - 1

Complete the diagram below to illustrate the shell model for argon as described in the preceding *Information* section. Each shell outside the +18 nuclus is indicated by a solid line. Label the shells (n,l) = (1,0), (2,0), (2,1), (3,0), and (3,1). Add the 18 electrons as small dark circles. Put 2 in each of the shells that have l = 0, and put 6 in each of the shells that have l = 1.

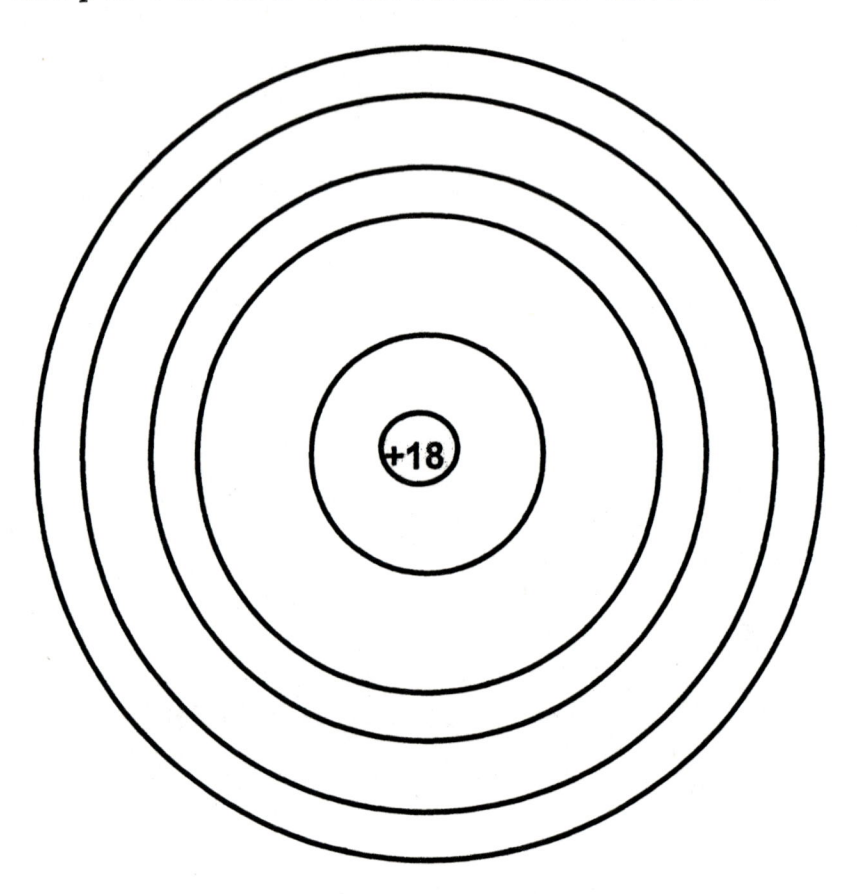

Research

Use your text as a resource and explain why there are 2 and only 2 electrons in each of the shells with l = 0, and 6 electrons in each of the shells with l = 1.

What do you predict?

Potassium has 19 protons and 19 electrons. The additional electron, compared to argon, goes in the n = 4, l = 0 shell.

1. Will the radius of a potassium atom be larger or smaller than that of argon? Explain in terms of the shell model diagram that you drew in *Application – 1* above.

2. Will it require more or less energy to remove an electron from potassium compared to argon? Explain in terms of the shell model diagram that you drew in *Application – 1* above. Remember, an electron is bound in an atom because it is attracted to the positively charged nucleus even though it is repelled by the other electrons, and the net result is a balance between the attraction by the nucleus and the repulsion by the other electrons.

Application - 2

Based on your experience with analyzing the argon photoelectron spectrum, construct a photoelectron spectrum, an energy level diagram, and a shell model for krypton. The atomic number of krypton is 36, and the electron binding energies are 14326, 1921, 1704, 293, 218, 94.5, 27.5, 14.1 eV.

General Chemistry: Guided Explorations

07-2. Periodic Trends in Properties of Elements

With knowledge of the shell structure of atoms and electron configurations, you can understand why different atoms are in the positions they are in the Periodic Table. You also can explain trends in some properties of the elements. In this activity, you examine trends in atomic radius, ionization energy, and electron affinity.

Exploration – 1: The Size of Atoms

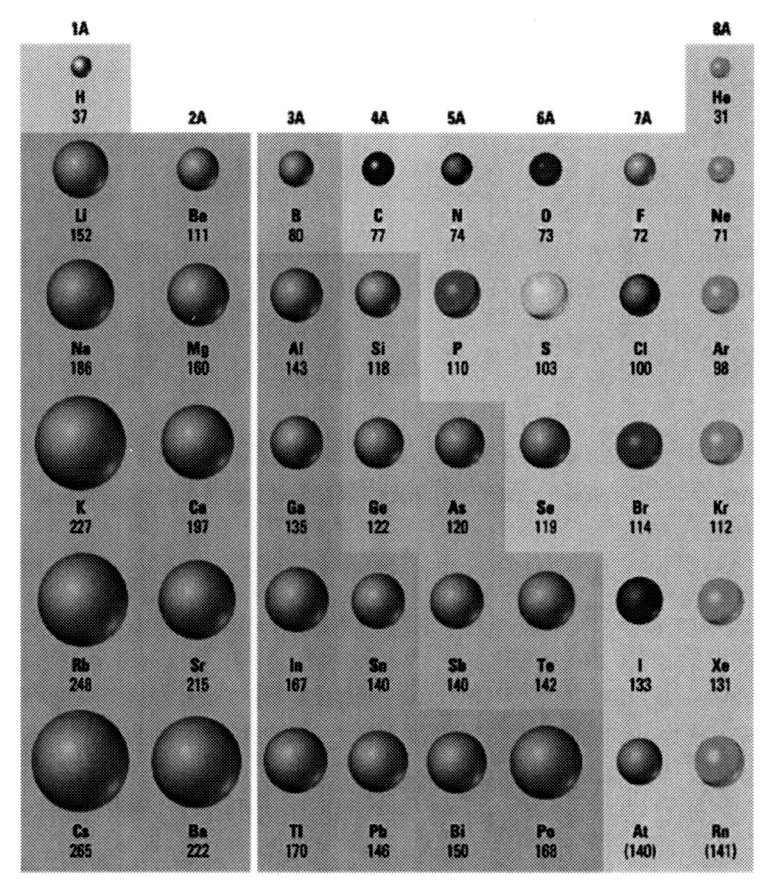

Fig. 1. Atomic radii of the main group elements in pm (10^{-12} m).

1.1. According to the information in Figure 1, does the atomic radius increase or decrease as the atomic number increases across a period (row) in the Periodic Table?

1.2. In going from one element to the next across a period, does the number of protons and the number of electrons increase or decrease?

1.3. An increase in the number of protons in the nucleus should increase the electron − nucleus attraction, pull the electrons in, and make the atom smaller. An increase in the number of electrons should increase the electron − electron repulsion, push the atoms apart, and make the atom larger. In view of you answer to Question 1.1, which effect evidently dominates in going from one atom to the next across a period: electron − nucleus attraction or electron − electron repulsion? Explain.

1.4. According to the pictures and values in Figure 1, what happens to the atomic radius down a group (column) in the Periodic Table?

1.5. In going from one element to the next down a group, what happens to the number of protons and the number of electrons?

1.6. In view of you answer to Question 1.4, which effect evidently dominates in going from one atom to the next down a group: electron − nucleus attraction or electron − electron repulsion? Explain.

1.7. Use the shell structure of atoms to explain why electron − nucleus attraction dominates and makes atoms smaller in going from one atom to the next across a period.

General Chemistry: Guided Explorations

1.8. Use the shell structure of atoms to explain why electron-electron repulsion dominates and makes atoms larger in going from one atom to the next down a group. Include in your explanation the fact that the distance of the outer electrons from the nucleus increases as one goes down a group.

Exploration – 2: Ionization Energies

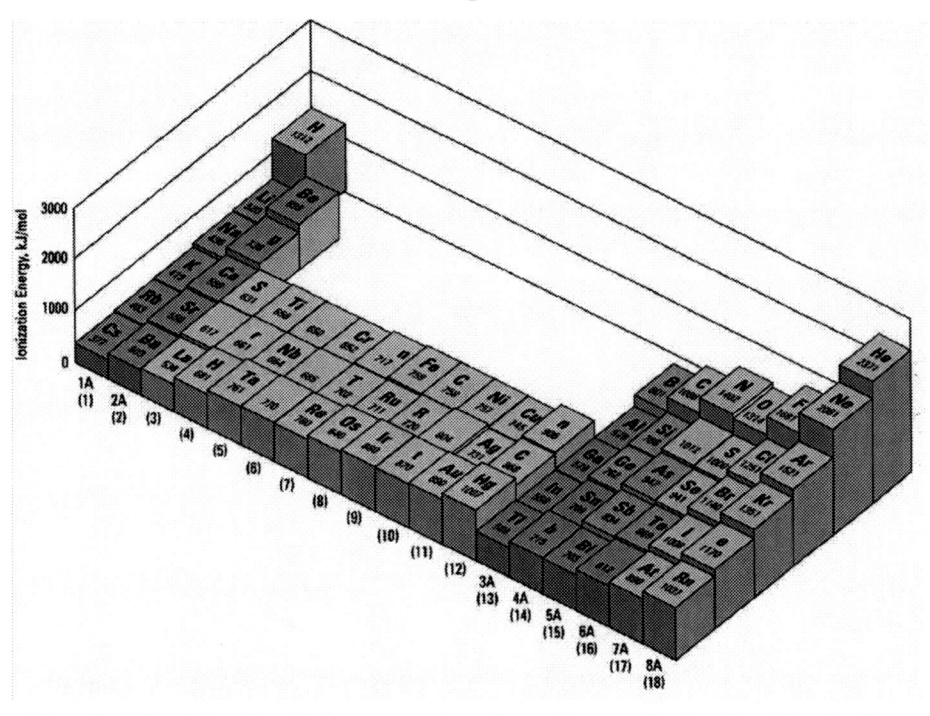

Fig. 2. First ionization energies for elements in the first six periods.

2.1. According to the information in Figure 2, does the ionization energy generally increase or decrease across a period in the Periodic Table?

65

2.2. Explain whether or not the changes in ionization energy and atomic radii across a period are consistent with the following statement: an increase in the electron-nucleus attraction increases the energy required to remove an electron from an atom and pulls electrons closer to the nucleus.

2.3. According to the information in Figure 2, does the ionization energy increase or decrease down a group in the periodic table?

2.4. Explain whether or not the changes in atomic radii and ionization energies down a group are consistent with the following statement: an increase in the distance of an electron from the nucleus and an increase in electron-electron repulsion decreases the energy required to remove an electron from an atom.

Application

Using only a Periodic Table and your knowledge of periodic trends in atomic size and ionization energies that you gained by answering the above questions, make the following predictions.

1. For each pair, predict which one is larger:
   N or O, H or He, Br or Cl, Na or Mg, Si or C, Ge or P

2. For each pair, predict which one has the higher ionization energy: H or He, Ne or Ar, Na or Ne, F or Cl, K or Cs

General Chemistry: Guided Explorations

Exploration – 3: Electron Affinities

The *ionization energy* of an atom is defined as the energy required to ionize an atom, corresponding to the following reaction equation.

$$X \rightarrow X^+ + e^- \qquad (1)$$

By analogy, research scientists define the *electron affinity* of an atom as the energy required to ionize the atomic anion, corresponding to the following reaction equation. The electron affinity then is a positive quantity.

$$X^- \rightarrow X + e^- \qquad (2)$$

In some textbooks, the electron affinity is defined as the energy associated with the following electron attachment reaction.

$$X + e^- \rightarrow X^- \qquad (3)$$

Because electron attachment, Equation 3, is just the reverse of anion ionization, Equation 2, the textbook convention produces electron affinities with the same magnitude but with a negative sign compared to the electron affinities used by research scientists and listed in the table below.

Electron Affinities (kJ/mol)

1A (1)	2A (2)	3A (13)	4A (14)	5A (15)	6A (16)	7A (17)	8A (18)
H 73							He 0
Li 60	Be 0	B 27	C 122	N 0	O 141	F 328	Ne 0
Na 53	Mg 0	Al 43	Si 134	P 72	S 200	Cl 349	Ar 0
K 48	Ca 2	Ga 30	Ge 119	As 78	Se 195	Br 325	Kr 0
Rb 47	Sr 5	In 30	Sn 107	Sb 103	Te 190	I 295	Xe 0

Fig. 3. Electron affinities of main group elements.

3.1. In each group, which element has the largest electron affinity?

3.2. In each group, which elements have an electronegativity of 0 or near 0?

3.3. Draw a shell model diagram for Ne and Ne⁻, and use the diagram to explain why the electron affinities of the noble gases are all 0. In your explanation include the effects of the distance of the outer electron from the nucleus, electron-electron repulsion, and electron-nucleus attraction.

3.4. Draw a shell model diagram for F and F-, and use the diagram to explain why the electron affinities of the halogens are all very large. In your explanation include the effects of the distance of the outer electron from the nucleus, electron-electron repulsion, and electron-nucleus attraction.

3.5. In the second period, the electron affinities of beryllium and nitrogen are 0. The electron configuration of Be⁻ is $1s^2 2s^2 2p^1$ and that of N⁻ is $1s^2 2s^2 2p_x^2 2p_y^1 2p_z^1$. Compare these configurations and shell model diagrams with those of Be and N, and in terms of electron-electron repulsion, explain why Be⁻ is not stable compared to Be, and N⁻ is not stable compared to N.

General Chemistry: Guided Explorations

08-1. Lewis Structures: What strategy produces the correct Lewis structure every time?

Atoms combine to form molecules because molecules are more stable than atoms. *More stable* means that molecules have a lower energy than the separated atoms.

The attractive force that holds a pair of atoms together in a molecule is called a *bond*. One picture of bonding views a bond as a pair or electrons between two nuclei. This simple model works very well. The negative charge of the two electrons attracts each nucleus, thereby effectively attracting the two positively charged nuclei to each other. This attraction holds the two atoms together.

Bonds are formed by the outer least tightly bound electrons. These electrons are called the *valence electrons*. The valence electrons generally come from the highest energy s and p orbitals.

G.N. Lewis figured out a way to determine how the valence electrons in a molecule arrange themselves to form bonds between atoms. Lewis reasoned that since the noble gases are very stable and have 8 electrons (*an octet*) in their outermost or valence shell, then stable molecules would be produced if one or more pairs of electrons were shared between the bonded atoms to give each atom an octet of electrons.

Bond formed by atoms sharing pairs of electrons are called *covalent bonds*. Lewis developed a system of drawing diagrams to determine the arrangement of pairs of electrons in molecules. This arrangement is called the *electron structure* or *electron-pair geometry* of the molecule, and the diagrams are called *Lewis structures*.

In this activity, you will learn how to draw Lewis structures, determine how many pairs of nonbonding electrons there are in a molecule, and identify whether the bonds are single, double, or triple covalent bonds.

Figure 1 provides examples of Lewis structures of three molecules showing a carbon-carbon triple bond, a carbon-oxygen double bond, and lots of nonbonding electron pairs.

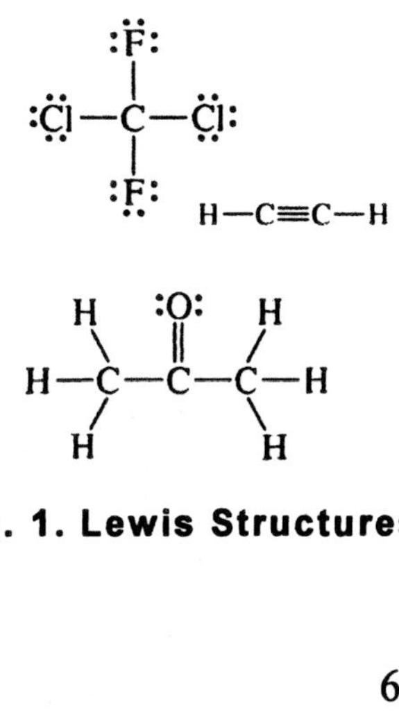

Fig. 1. Lewis Structures

Exploration – A Strategy for Drawing Lewis Structures

Step	Examples
1. Count the total number of valence electrons in the molecule or ion. The Group number (IA – VIIIA) in the column heading of the Periodic Table gives you the number of valence electrons for a neutral atom. Subtract the ion charge for a cation, add the ion charge for an anion.	PCl_3 has 26 valence electrons: 5 from P and 3 x 7 = 21 from Cl. NO_2^- has 18 valence electrons: 5 from N, 2 x 6 = 12 from O, and 1 more because it is a -1 anion.
2. Decide which atoms are bonded to each other, and use a pair of electrons, represented by a line to connect the atomic symbols for these atoms. The central atom in a structure usually is the one written first in the molecular formula and can form the most bonds with other atoms.	Cl—P—Cl Cl O—N—O
3. Place lone pairs (aka nonbonding pairs) of electrons around each atom to satisfy the octet rule starting with the atoms on the outside. H only gets 2 electrons.	:Cl—P—Cl: :Cl: :O—N—O:
4. Place any leftover electrons on the central atom even if it will give that atom more than an octet.	:F: :F—S—F: :F:
5. If the number of electrons around the central atom is less than 8, use one or more lone pairs from an outside atom to form double or triple bonds until all the atoms have an octet of electrons. A double bond has 2 shared pairs, and a triple bond has 3 shared pairs.	O=N—O:
6. Draw any equivalent resonance structures. The two bonds between N and O in the nitrite anion are equivalent. To represent this fact, it is necessary to draw two Lewis structures, and nitrite is said to be a resonance hybrid of these two structures.	O=N—O: :O—N=O

7. Calculate and write the formal charge on each atom. For each atom, the formal charge = the number of valance electrons contributed by the atom in Step 1 – the number of lone pair electrons – half the number of bonding electrons.	0 0 0
	:Cl—P—Cl:
	    :Cl:
	     0

1. How can you determine the number of valence electrons that an atom has?

2. Given a molecular formula, which atom usually is the central atom in the Lewis structure?

3. What do you do in drawing a Lewis structure if each atom has been assigned an octet of electrons, but you have not used all the valence electrons?

4. What do you do in drawing a Lewis structure if you have used all the valence electrons, but each atom does not have an octet of electrons?

5. Why is it necessary to draw equivalent resonance structures to represent a molecule?

6. How is the formal charge of an atom determined?

7. How can the formal charge be used to determine which of 2 possible Lewis structures has the lower energy and therefore is the more stable? The Lewis structure with the lower energy is the more likely or reasonable structure for the ground state of the molecule. Use your textbook as a resource to answer this question.

Application

1. Draw Lewis structures for the following molecules. Include formal charges in your diagrams.

    hydrofluoric acid, HF

    ammonia, NH_3

    phosphate, PO_4^{3-}

    butane, C_4H_{10}

    ethanol, CH_3CH_2OH

    carbon dioxide, CO_2

    acetylene, C_2H_2

    Chloroethene, C_2H_3Cl

    Chlorine trifluoride, ClF_3

    nitrate, NO_3^-

2. Draw Lewis structures for carbonyl sulfide with three possible connectivities: O-C-S, C-O-S, and C-S-O. Use the formal charges on the atoms as a criterion to predict the structure of this molecule.

Got It!

Use Lewis structures to predict whether thiocyanide has the structure S-C≡N or S-N≡C. Note thiocyanide has a -1 charge.

08-2 Electronegativity and Bond Properties: Which atom attracts more electron density?

In covalent bonds, pairs of electrons are shared between two atoms, but the electrons are not shared equally unless the two atoms are the same. In molecules such as oxygen and nitrogen, O_2 and N_2, where both atoms are the same and the sharing is equal, the bond is said to be *nonpolar*. When the atoms are different and the electrons are pulled more toward one atom than the other, the bond is said to be *polar*. It even is possible for one atom to pull the valence electrons away from the other completely to form ions. In that case the bond is said to be *ionic*.

Electronegativity is the ability of an atom to attract electrons away from another atom in a molecule. Linus Pauling first proposed the concept of electronegativity and assigned an electronegativity value to each atom, giving fluorine, which is the most electronegative atom, a value of 4.0. Electronegativity can be thought of as an average of the *ionization energy* and *electron affinity* of an atom. The ionization energy is a measure of the strength of an atom in keeping an electron, and the electron affinity is a measure of the strength of an atom in attracting another electron. If both of these are large, then the electronegativity is high. If both of these are small, then the electronegativity is low.

Exploration – 1. Electronegativity of the Elements

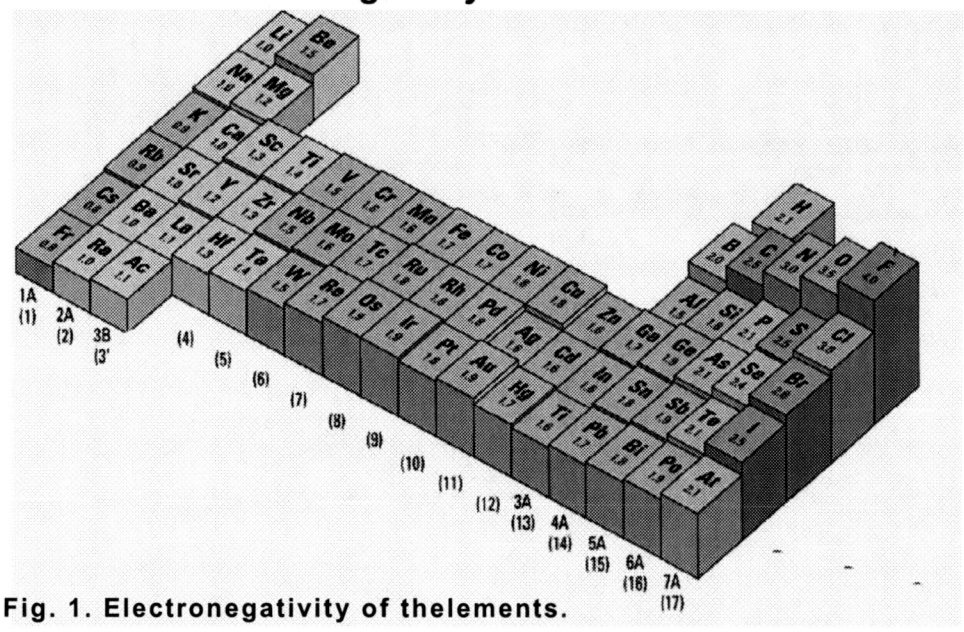

Fig. 1. Electronegativity of the lements.

1.1. Compare the ionization energies and electron affinities in Activity 07-2 with Figue 1 above. In which part of the Periodic Table are elements with high ionization energies, high electron affinities, and high electronegativities found?

1.2. In which part of the Periodic Table are elements with low ionization energies, low electron affinities, and low electronegativities found?

1.3. Which are least electronegative and which are the most electronegative, metals or nonmetals?

1.4. The polarity of a bond is quantified by a value called the dipole moment, μ. The larger the dipole moment, the more polar the bond. What relationship between bond polarity and the electronegativity (EN) of the elements can you identify from the data in Table 1?

Table 1. Measures of Bond Polarity

Bond	μ in debye	EN values	Δ(EN)
F – F	0	F = 4.0	0
H – Br	0.8	H = 2.1 Br = 2.8	0.7
H – F	1.9	H = 2.1 F = 4.0	1.9
Na – F	8.2	Na = 0.9 F = 4.0	3.1

Information

A larger difference in electronegativity between two atoms produces a more polar the bond. The shift from nonpolar covalent bonds to ionic bonds can be regarded as a continuum as shown in Figure 2. Polar covalent bonds are in the middle, and slightly and very polar covalent bonds are on either side of the middle. Ions and an ionic bond generally are formed when the electronegativity difference between the atoms is 2.0 or larger.

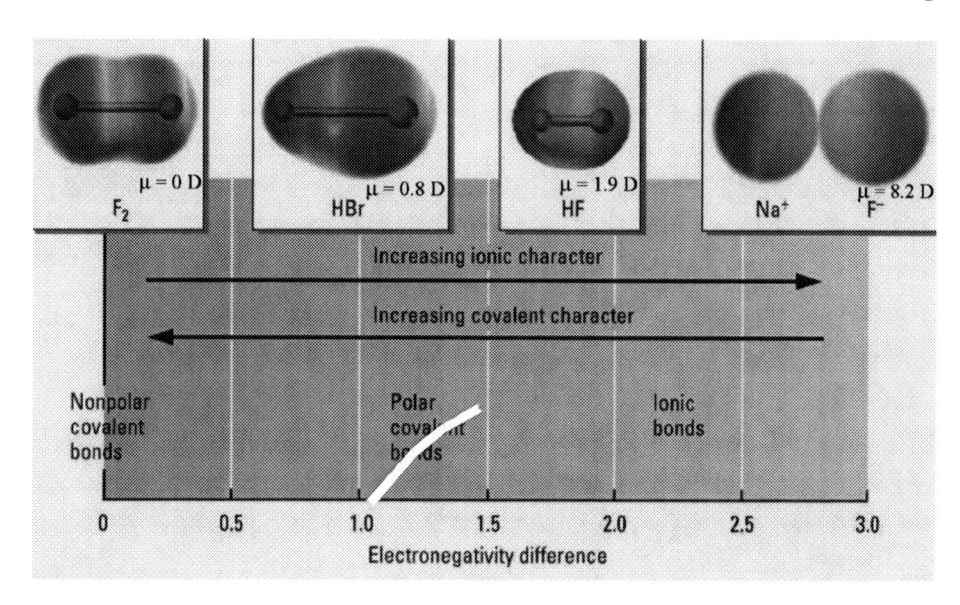

Fig. 2. Bond character, dipole moments, and electronegativity differences for four substances.

Got It!

For each of the following pairs of bonds, identify which atom is partially positive and which is partially negative, and indicate which bond is the more polar.

1.1. Cl – F and Cl – Br

1.2 Si – H and C – H

1.3. P – Cl and S – Cl

1.4. C – O and C – N

1.5. C – O and Si – O

Exploration – 2. Bond Lengths and Strengths

Table 2. Bond Lengths and Strengths

Bond	Bond Length (pm)	Bond Strength (kJ/mol)
C – C	154	356
C = C	134	598
C ≡ C	121	813
N – N	140	160
N = N	120	418
N ≡ N	110	946

2.1. As the multiplicity of the bond increases from single to double to triple, what happens to the bond length?

2.2. As the multiplicity of the bond increases from single to double to triple, what happens to the bond strength?

2.3. Does the electron density between the two nuclei increase or decrease as the bond multiplicity increases, and how does this change affect the bond length and bond strength?

2.4. Why are N – N bond lengths shorter than C – C bond lengths? Explain in terms of the relative sizes of N and C atoms, see Activity 07-2 for information on atomic size.

Got It!

For each of the following pairs of bonds, identify the pair that you expect to be shorter and have the higher bond energy. Explain your rationale in each case.

2.1. Si – O or C – O

2.2. C = O or C = C

2.3. C – Cl or C – Br

2.4. N ≡ O or N = O

09-1 VSEPR Model: How can the geometry or shape of a molecule be predicted?

The geometry or shape of a molecule is important in determining its reactivity, its function in biological systems, and the physical and chemical of the substance. The valence-shell electron-pair repulsion model (VSEPR) provides a simple and reliable method for predicting the shapes of molecules and polyatomic ions. The VSEPR model is based on the idea that regions of high electron density in a molecule repel each other. These regions include the bonding electron pairs and the nonbonding lone pairs of electrons.

Exploration – A Strategy for Determining Molecular Shapes

Step	Examples
1. Draw the Lewis structure. See Activity 08-1.	H—N̈—H with H below
2. Count the number of bonds and lone pairs around each atom. Count double and triple bonds as 1.	In ammonia: 3 bonds + 1 lone pair = 4
3. Arrange the bonds and lone pairs to maximize their separation, which minimizes the electron – electron repulsion energy.	Four regions of electron density are furthest apart if they have a tetrahedral shape. Electron-pair geometry (tetrahedral)
4. Add the atoms to positions that are consistent with the Lewis structure, and keep the lone pairs (if there is more than one lone pair) as far apart as possible.	
5. Determine the molecular shape from the positions of the atoms, ignoring the lone pairs.	Ammonia is a triangular pyramid.

1. In Step 1, you need to determine the Lewis structure. Write below the procedure that you follow to draw a correct Lewis structure.

2. In determining the electron pair geometry, why are the bonding and lone pair electrons spaced as far apart as possible?

3. According to Step 4, if you have two lone pairs and four atoms around a central atom, would you position the lone pairs at 90° or 180° to each other? Explain.

4. There are 6 possible arrangements or geometries for the bonds and lone pairs of electrons. Use your textbook as a reference to help you complete Table 1, below.

Table 1. Electron Pair Geometries

Number of bonds and lone pairs.	Electron Geometry	Sketch of the Electron Pair Geometry
2	Linear	
3	Triangular planar	
4	Tetrahedral	
5	Triangular bipyramidal	
6	Octahedral	

5. In Step 4 you add the atoms to the electron pair geometry, and in Step 5, you determine the molecular shape by looking at the positions of the atoms and ignoring the lone pairs. Sketch the molecular shape for the following examples..

(a) 3 atoms, linear, CO_2

(b) 3 atoms, bent or angular, SO_2

(c) 4 atoms, triangular planar, CO_3^{2-}

(d) 4 atoms, triangular pyramidal, NH_3

(e) 4 atoms, T shaped, ClF_3

(f) 5 atoms, tetrahedral, CH_4

(g) 5 atoms, seesaw, SF_4

(h) 5 atoms, square planar, XeF_4

(i) 6 atoms, square pyramidal. BrF_5

(j) 6 atoms, triangular bipyramidal, PF_5

(k) 7 atoms, octahedral, SF_6

Got It!

For the following molecules: draw the Lewis structure, and; use the VSEPR model to identify the electron pair geometry and the molecular shape.

Molecule	Lewis Structure	Electron Pair Geometry	Molecular Shape
H_2O			
I_3^-			
$SeCl_6$			
SbF_5			
H_2CO			
SeO_3^{2-}			
SiF_4			
KrF_4			
SF_4			
ICl_3			
BrF_5			

09-2. Hybrid Atomic Orbitals: How can atomic orbitals describe all the different electron geometries found in molecules?

Review of Prerequisite Knowledge

The forces that hold molecules together are electrostatic. It is as simple as opposite charges (+ and -) attract and like charges (+ and + or - and -) repel. However, when the theory of electrostatics was applied, it could not account for any of the properties of atoms and molecules or even for their existence. Classical electrostatic theory says that a proton and an electron should attract each other, come together, and neutralize each other, not exist as a hydrogen atom.

As a result of this situation, a new theory was invented. This theory is known as *Quantum Mechanics* or *Schrödinger's Wave Mechanics*. Schrödinger figured out how to find the waves that describe electrons in atoms and molecules. Application of Schrödinger's ideas to the hydrogen atom, produced the wave functions Ψ_{nlm} for the hydrogen atom. Each wave function is identified by unique values for the quantum numbers n, l, and m_l. Any question about the hydrogen atom can be answered by using its wave functions. These functions are called atomic orbitals. Pictures of these functions show the regions of space where the function has large positive or negative values.

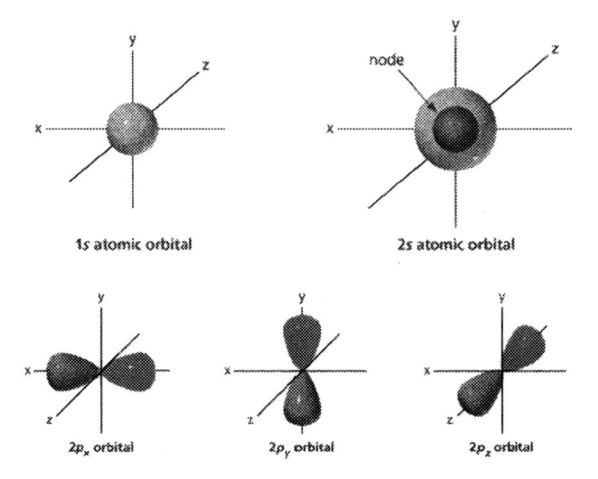

Fig. 1 Pictures of the 1s, 2s, and 2p atomic orbitals.

The square of the wave function Ψ_{nlm}^2 is the electron probability density. It gives the probability density of finding an electron at different points in space.

In a chemical bond there is a high density of electrons between the two atomic nuclei. It is the negative charge from the electrons that pulls the two positively charged nuclei together to form the bond. However, the atomic p-orbitals don't describe many of these bonds very well because they point in the wrong directions. Consider the case of methane shown below.

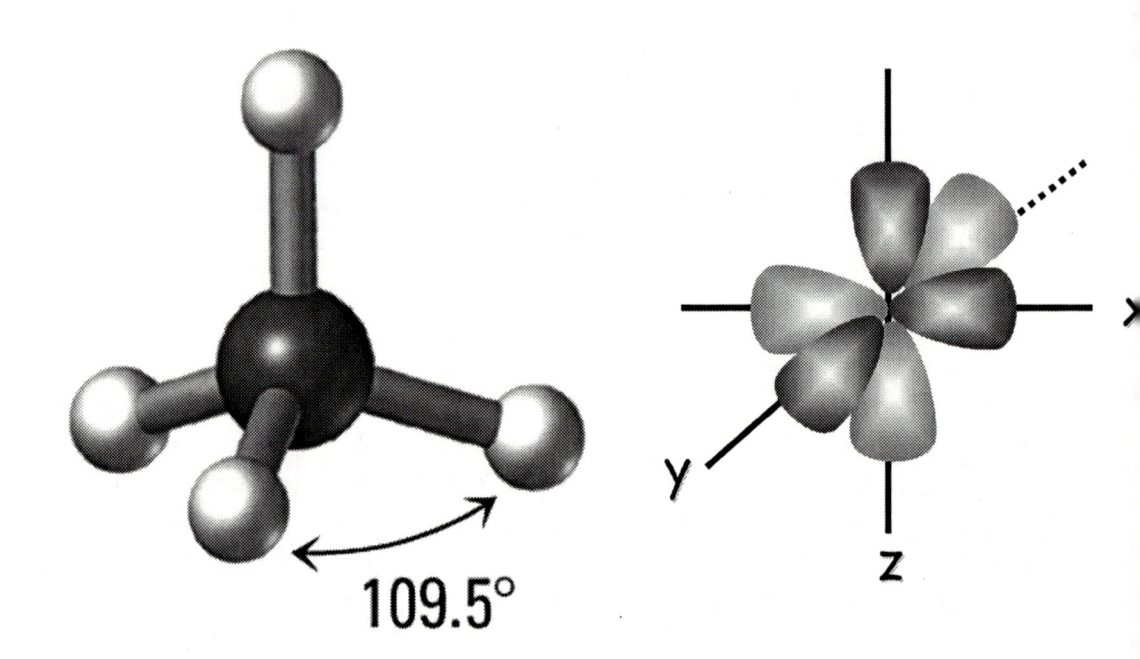

Fig. 2 Methane CH₄ and the three atomic p-orbitals.

Figure 2 shows that methane has a tetrahedral shape, and that the three atomic p orbitals of carbon are directed along the x, y, and z axes in both directions. There is a high electron density between the C and H atoms in methane, but the atomic p-orbitals do not lie between the atoms as can be seen in Figure 2. The atomic p-orbitals are off to one side of the bonds or the other and extend in both directions along the x, y, and z axes. The atomic p-orbitals do not describe where the electrons are in the methane molecule.

New orbitals are needed. These new orbitals are formed by combining the s and p orbitals   The orbitals made from this combination are called *hybrid orbitals*. The hybrid orbitals are designed to lie along the bonds.

Five different types of hybrids can be made using combinations of s, p, and d orbitals. These five types of hybrid orbitals are shown in Figure 3.

General Chemistry: Guided Explorations

Exploration: Hybridization of Atomic Orbitals

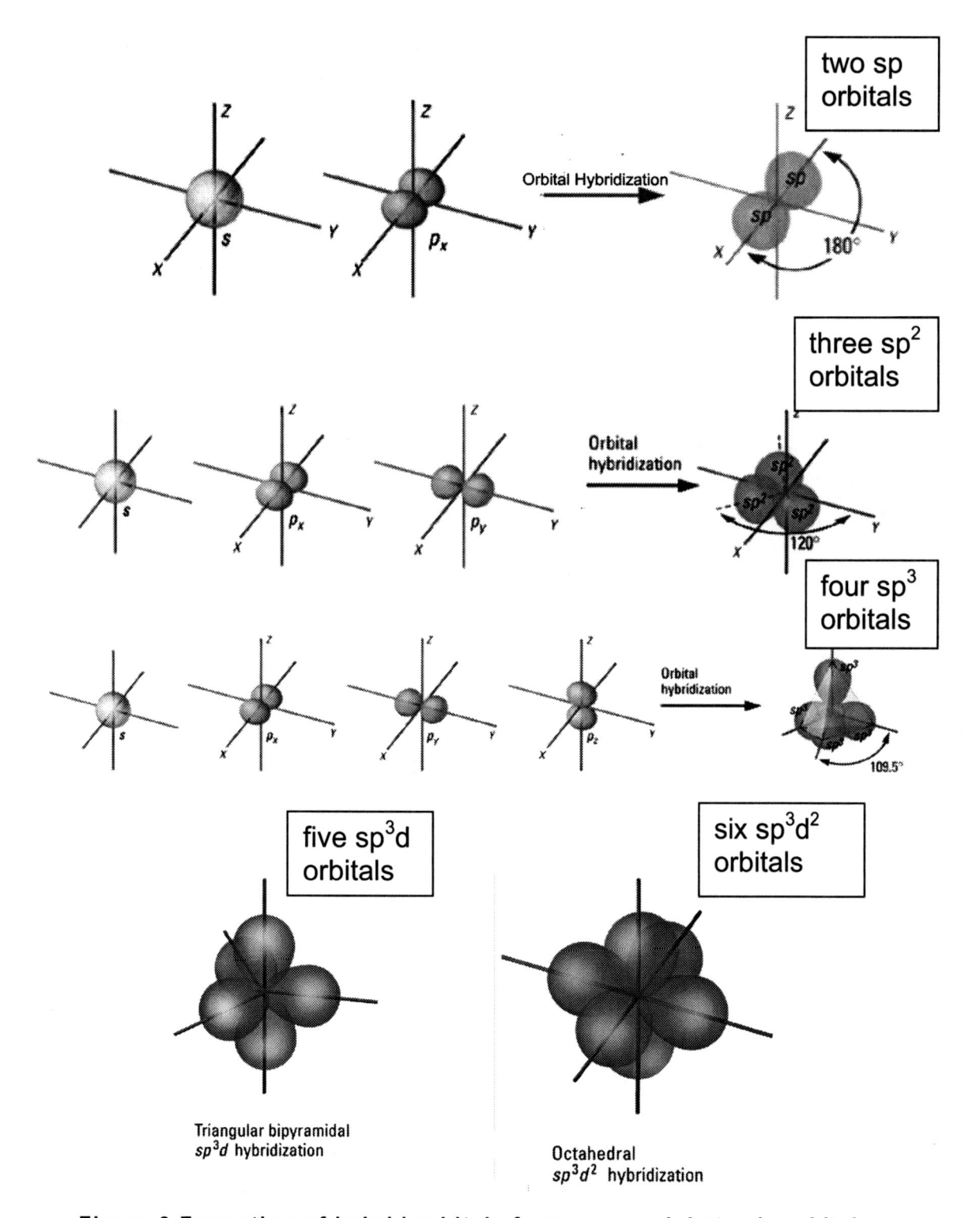

Figure 3 Formation of hybrid orbitals from s, p, and d atomic orbitals.

83

1. What are the names of the five types of hybrid orbitals shown in Figure 3? For example one type is sp³.

2. How many different hybrid orbitals are there of each type? For example, there are four sp³ orbitals.

3. How many atomic orbitals and which ones must be combined to make each type of hybrid orbital? For example, four atomic orbitals (s, p$_x$, p$_y$, and p$_z$) combine to produce the four sp³ hybrid orbitals.

4. In general, what is the relationship between the number of atomic orbitals used and the number of hybrid orbitals produced?

5. What is the angle between the hybrid orbitals for each type? For example in sp³, the angle between the hybrid orbitals is 109°.

General Chemistry: Guided Explorations

6. Which set of hybrid orbitals in Figure 3 match the bond angles shown in Figure 2 for methane?

7. Figure 4 shows the hybrid orbitals used to describe the bonds in acetylene C_2H_2. Which set of hybrid orbitals in Figure 3 matches the bond angles in acetylene?

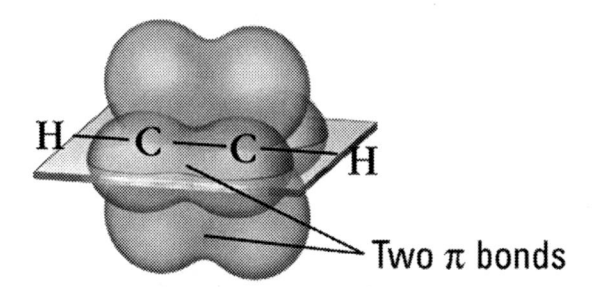

Fig.4 Bonding in acetylene. The lines represent bonds formed using carbon sp hybrid orbitals. The "clouds" represent bonds formed from the carbon atomic p_y and p_z orbitals on the two carbon atoms. The pair of p_y orbitals form one pi bond, and the pair of p_z orbitals form a second pi bond.

8. Figure 5 shows the hybrid orbitals used to describe the bonds in ethylene $CH_2=CH_2$. Which set of hybrid orbitals in Figure 3 matches the bond angles in ethylene?

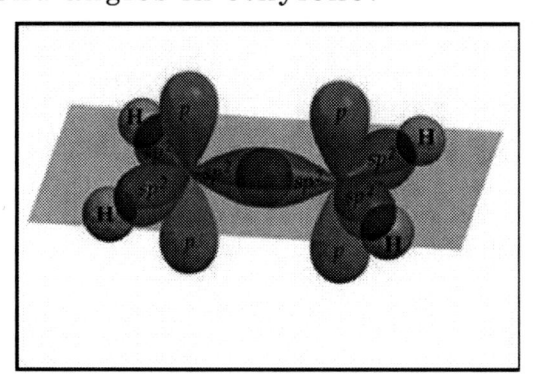

Fig. 5 Atomic and hybrid orbitals for ethylene. Note the p_z orbitals on the two carbon atoms overlap and form a pi bond just as they do in acetylene.

9.. Why is it necessary to include the concept of hybrid orbitals in the description of molecules?

Information

Molecular orbitals and bonds are identified as being either sigma or pi using the Greek symbols σ and π, repectively. Sigma orbitals and bonds are formed from the overlap of atomic or hybrid orbitals along the axis or line connecting the two atomic nuclei. Pi orbitals and bonds are formed from the sideways overlap of parallel atomic p orbitals.

Application

1. List the number of sigma and pi bonds for each one of the following: methane, ethylene, and acetylene.

Compound	Sigma Bonds	Pi Bonds
Methane, CH_4		
Ethylene, C_2H_4		
Acetylene, C_2H_2		

2. Using your knowledge of the Lewis structure and the VSEPR model, determine the geometry or shape of each of the following molecules. Identify the hybrid orbitals of the central atom in each, and give the expected bond angles.

   carbon tetrafluoride, CF_4 carbonate anion, CO_3^{2-}

   ozone, O_3 sulfur dioxide, SO_2

   formaldehyde, H_2CO nitrate anion, NO_3^-

Got It!

1. Identify the geometry and hybridization of each of the following phosphorous - chlorine species: PCl_4^+, PCl_5, PCl_6^-.

2. Identify the hybridization of each carbon atom in $CH_3CHCHCCH$. In this condensed structural formula, the H atoms are bonded to the carbon on their left. Draw a picture representing the shape of this molecule.

10-1. Gases and the Ideal Gas Law: How much air does it take to fill a balloon?

What do you predict?

Identify whether you *agree* or *disagree* with the following predictions pertaining to the gas inside a can of hair spray. Assume all the liquid has been used, and only some residual gas consisting of hair spray and propellant molecules remains.

1. If the temperature changes from 15 °C to 30 °C, the pressure inside the can will double.

2. If half the mass of the gas is removed from the can, the pressure inside the can will decrease by a factor of 2.

3. The can contains fewer molecules than an identical can of hydrogen gas at the same temperature and pressure because the hair spray molecules and propellant molecules are larger than hydrogen molecules.

Information

The relationships among the pressure (P), volume (V), Kelvin temperature (T), and amount (n moles) of a gas are expressed in a single equation called the *Ideal Gas Law*. This equation includes the gas constant (R).

$$PV = nRT \qquad (1)$$

This equation can be used to calculate P, V, n, or T whenever three of these four are known. The Ideal Gas Law is a good approximation for real gases at low pressures (a few atmospheres and less) and high temperatures (relative to the boiling point).

The Ideal Gas Law assumes that atoms and molecules are points with no volume, and that they do not interact except through collisions. These assumptions are good when the density is low so the atoms and molecules are far apart, and the temperature is high so the interaction energy is small compared to the kinetic energy of the atoms and molecules.

When you look at Equation (1), you should have several questions about it. The *Exploration* below will help you address these questions, understand what the equation is telling you, and identify how you can use it.

Exploration

1. What units must be used for volume, pressure, amount of substance, and temperature in the Ideal Gas Law? When used with the Ideal Gas Law, the gas constant R has units of L atm mol^{-1} K^{-1}.

2. What is the numerical value for the gas constant, R? Using the following information obtained from experimental measurements on many gases, calculate a numerical value for the gas constant R. At 273 K and 1 atm pressure, 1 mole of these gases occupied 22.4 L. This temperature and pressure are called the standard temperature and pressure for gases (STP), and 22.4 L is called the standard molar volume, which was found to be the same for all gases.

3. What predictions can be made from the Ideal Gas Law, when the following changes occur?
   (a) The Kelvin temperature doubles while n and V remain unchanged.

   (b) The volume doubles while n and T remain unchanged.

   (c) Gas is added to a container, while T and P remain unchanged.

Application

1. A valve is opened between a 10 L tank containing a gas at 5 atm of pressure, and an empty 15 L tank. What will be the new pressure of the gas, assuming the temperature does not change?

2. In Havre, Montana the temperature went from -34 °F on a day in January to 103 °F on a day in July in 2005. If you inflated your automobile tires to a pressure of 30 lb/in² on that day in January, what would the pressure be on that day in July, using lb/in²? Assume the volume of the tire and the amount of air in the tire did not change.

3. A 1.00 g sample of water vaporizes completely inside a 10.0 L container at 150 °C. What is the pressure of the water vapor?

4. A weather balloon is filled with helium to a volume of 1000. L on the ground where the pressure is 745 torr and the temperature is 20°C.. When the balloon ascends to a height of 2 miles, the pressure is 600 torr and the temperature is −33°C. What is the volume of the helium in the balloon at this altitude?

Got It!

1. You've *Got It!* if you can answer this question without doing any numerical calculations. Which one of the following gas samples contains the largest number of molecules, and which one contains the smallest? Explain your reasoning.
   (a) 1.0 L of H_2 a STP
   (b) 1.0 L of N_2 at STP
   (c) 1.0 L of H_2 at 27 °C and 760 torr.
   (d) 1.0 L of CO_2 at 0 °C and 800 torr.

2. If a gas in a jar with a leaky seal (P and V constant) is heated and the Kelvin temperature quadruples, what fraction of the gas remains in the jar, and what fraction of the gas escaped?

3. Reconsider the predictions that you made at the beginning of this activity. If you decide to change any of your predictions in view of what you learned from this activity, identify which ones and explain why.

10-2. Partial Pressures: What determines the total pressure of a mixture of gases?

What do you think?

Suppose you have two 1.0 L containers at 300 K. One contains nitrogen gas, and the other contains oxygen gas. The pressure inside both containers is 1.0 atm. You transfer the contents of these two containers into a third as shown in the diagram below. Under these conditions, nitrogen and oxygen behave as ideal gases.

Container 1   Container 2   Container 3

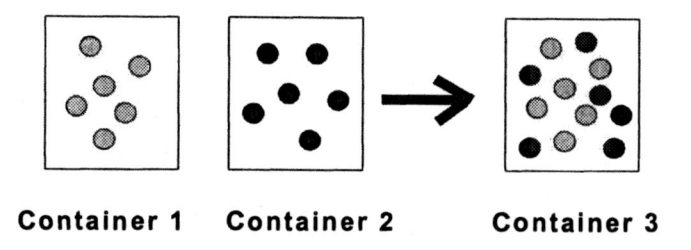

Fig. 1. Oxygen and nitrogen gases from separate containers are combined into one container. The containers are identical.

1. Has the pressure produced by oxygen changed?

2. Has the pressure produced by nitrogen changed?

3. What is the total pressure inside Container 3?

Exploration

1. For the situation described by Figure 1, determine the following items using the Ideal Gas Law..
   (a) The number of moles of oxygen in Container 1, and the number of moles of nitrogen in Container 2.

   (b) What is the total number of moles in Container 3?

(c) Show how you can calculate the total pressure of the gases in Container 3 by using the Ideal Gas Law.

(d) Show how you can calculate the pressure of oxygen in Container 3 by using the ideal gas law with n = moles of oxygen. This pressure is called the *partial pressure* of oxygen.

(e) Show how you can calculate the pressure of nitrogen in Container 3 by using the ideal gas law with n = moles of nitrogen. This pressure is called the *partial pressure* of nitrogen.

2. Use your result from Question 1 to calculate the ratio of the partial pressure of oxygen to the total pressure.

3. Use your result from Question 1 to calculate the ratio of the moles of oxygen to the total number of moles. This ratio is called the *mole fraction* of oxygen present in the mixture.

4. Are the values you calculated in Questions 2 and 3 equal or different? Divide the ideal gas law equation for the partial pressure of oxygen ($P_{O_2}V = n_{O_2}RT$) by the ideal gas law equation for the total pressure to show why the mole fraction must equal the ratio of partial pressure to total pressure.

General Chemistry: Guided Explorations

Information

John Dalton was the first to observe that the total pressure exerted by a mixture of gases is the sum of the partial pressures of the individual gases in the mixture. This statement is known as Dalton's law of partial pressures, which can be written in the form of an equation.

$$P_i = \sum_{sum\,over\,all\,gases} P_i, \text{ where } P_i \text{ is the partial pressure of gas } i \qquad (1)$$

This property is a consequence of the fact that ideal gas molecules act independently so the pressure of a gas inside a container depends on the number of molecules but not their identity. The pressure of one mole of oxygen plus one mole of nitrogen therefore will be the same as the pressure of two moles of nitrogen or two moles of oxygen.

This idea follows from the ideal gas law as shown below.

For a mixture of 3 gases

$$P_{total} = n_{total} \frac{RT}{V}$$

$$n_{total} = n_1 + n_2 + n_3$$

Substituting for n gives

$$P_{total} = (n_1 + n_2 + n_3)\frac{RT}{V}$$

$$P_{total} = \left(n_1 \frac{RT}{V} + n_2 \frac{RT}{V} + n_3 \frac{RT}{V}\right)$$

$$P_{total} = P_1 + P_2 + P_3$$

Dividing P_1 by P_{total} gives n_1/n_{total} because the RT/V factors cancel. So the partial pressure of a gas equals its mole fraction in a mixture times the total pressure of all the gases.

$$P_i = X_i P_{total}, \text{ where } X_i = \frac{n_i}{n_{total}}$$

Got It!

Mixtures of 15% oxygen and 85% helium are used in some scuba diving tanks to help prevent the bends, which is a condition caused by nitrogen bubbles forming in tissues and blood. A standard scuba tank has a volume of 18 L and can be filled to a pressure of 3000 psi. Assume the tanks are filled on a hot day when the temperature is 30°C.

1. What are the partial pressures of oxygen and helium in one of these tanks?

2. How many liters of oxygen at STP need to be used to fill one of these tanks?

3. How many liters of helium at STP need to be used to fill one of these tanks?

4. How many moles of oxygen and moles of helium are used in filling one of these tanks?

5. What is the total mass of the gases in one of these tanks when it is filled?

11-1. Phases of Matter: As water cools, does the temperature continue to drop as ice forms?

What do you think?

1. You have ice-cold Coke with ice in a glass ice sitting on a table on a very hot day. Is the temperature of the Coke increasing, decreasing, or staying the same as the ice melts?

2. You have water in a pot boiling on the stove. Is the temperature of the water increasing, decreasing, or staying the same?

3. When water boils, what is inside the bubbles that form?

Information

There are three states of matter (solid, liquid, and gas), each with distinct characteristics. A *state* of matter also is called a *phase*. The solid state of a substance, however, can have more than one phase. Each solid state phase has a different structure. For example, diamond and graphite are two solid state phases of carbon.

Diamond consists of a three dimensional network of covalently bonded carbon atoms. In diamond, each carbon atom is bonded to four others, and the atomic orbitals are sp^3 hybridized.

Graphite has layers of a two dimensional network of covalently bonded carbon atoms. In graphite, each carbon atom is bonded to three others, and the atomic orbitals are sp^2 hybridized.

Transitions between one phase and another can be caused by heating or cooling to change the temperature or by increasing or decreasing the pressure to change the spacing between the molecules.

Fig. 1.   Solid   Liquid   Gas

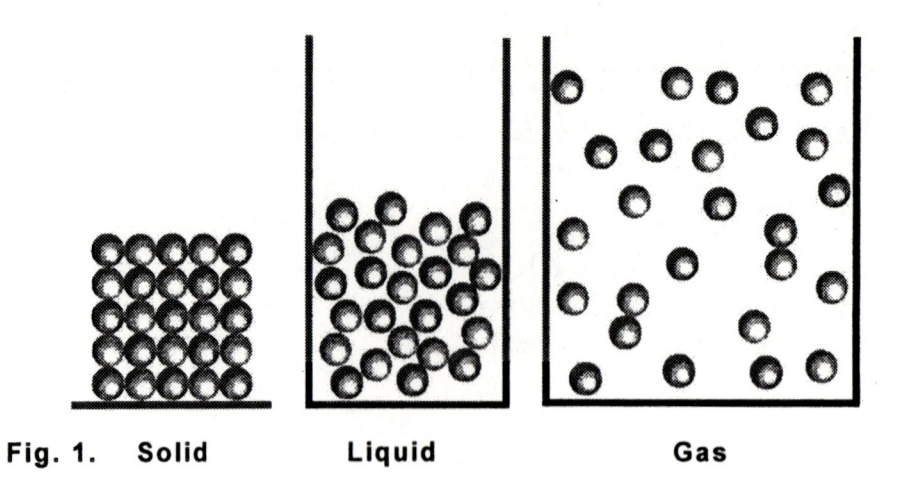

Table I. Characteristics of the States of Matter

Solid	Liquid	Gas
Long range order	Short range order	Disordered
Tightly packed	Closely spaced	Widely spaced
Vibrate about fixed position	Limited motion – rotate and diffuse	Free motion
Almost incompressible	Not very compressible	Highly compressible
Fixed shape	No fixed shape	No fixed shape
Does not flow	Can flow	Can flow

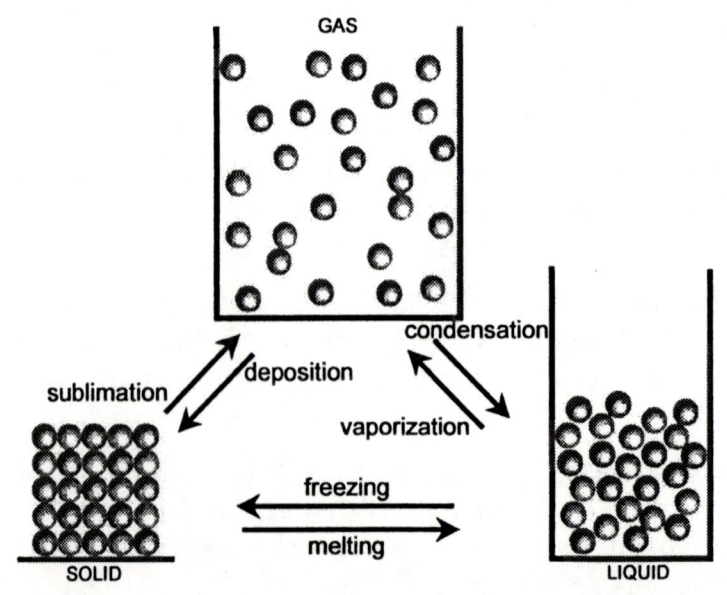

Fig. 2. The three states of matter and the six phase transitions between them.

General Chemistry: Guided Explorations

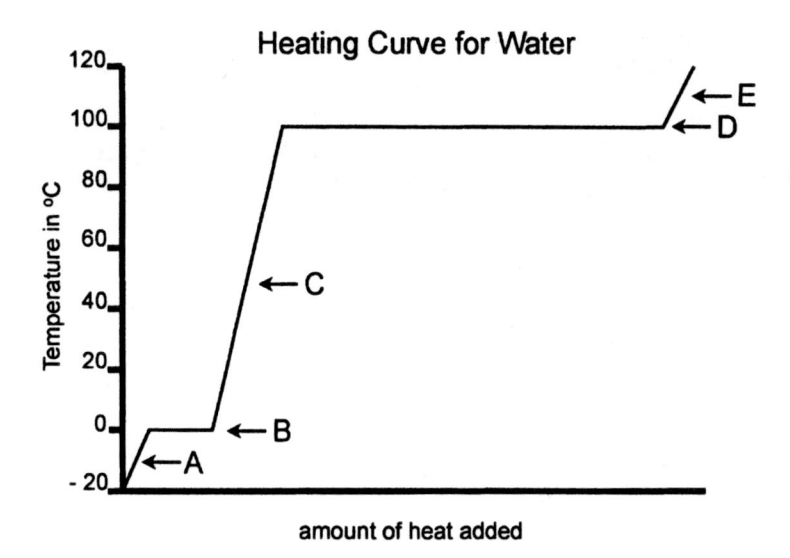

Fig. 3. Heating curve for water showing how the temperature changes as heat is added at a constant rate.

Exploration

1. In view of Figure 1 and the information in Table I, what do you see as the three most significant differences between the three states of matter?

2. What are the three processes listed in Figure 2 that convert one phase into another when a substance is heated?

3. What are the three processes listed in Figure 2 that convert one phase into another when a substance is cooled?

97

4. Figure 3 shows a heating curve for water. This data is obtained by measuring the temperature when heating at a constant rate. Five regions are identified in the figure and are labeled A through E. Below 0 °C, water is in the solid state (ice), between 0 and 100 °C water is in the liquid state, and above 100 °C water is in the gas state (steam).

   (a) Next to the labels A, C, and E in Figure 3, write the state of water (solid, liquid, gas) in that region of the graph.

   (b) What is happening to the water at the temperature labeled B in the figure?

   (c) What is happening to the water at the temperature labeled D in the figure?

   (d) Why does the temperature start to rise again at the points labeled B and D in the figure?

5. Based on the graph in Figure 3, do you think it takes more energy to melt 100 g of ice, raise the temperature of 100 g of water from 0 °C to 100 °C, or to boil and vaporize 100 g of liquid water? Explain.

Got It!

1. How can heating curves be used to identify the temperatures at which phase transitions occur for any given substance at any given pressure?

2. Reconsider your *What do you think?* responses at the beginning of this activity. If you decide to change any in view of what you have learned, identify which ones and explain why.

11-2. Phase Diagrams: What information can be obtained from a phase diagram?

A plot of the temperatures and pressures at which phase transitions occur is called a *phase diagram*. A generic phase diagram is shown below.

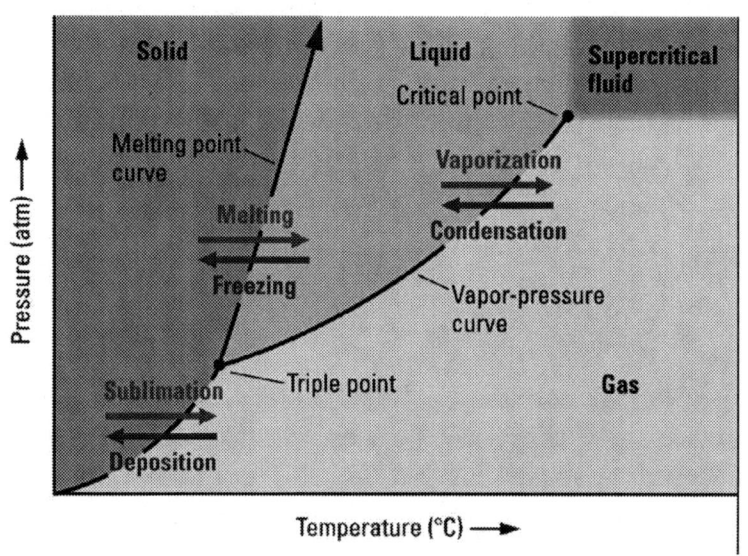

Fig. 1. A generic phase diagram showing the regions of temperature and pressure where the substance is a solid, liquid, and gas.

Exploration

1. In a phase diagram, areas represent different phases, and the boundary lines show the temperatures and pressures where two phases are in equilibrium and a phase transition occurs.
(a) What is the boundary line called in the Figure 1 where the liquid and gas are in equilibrium? What two phase transitions occur along this boundary line?

(b) What is the boundary line called in Figure 1 where the solid and liquid are in equilibrium? What two phase transitions occur along this boundary line?

(c) What label would you propose for the boundary line where the solid and the gas are in equilibrium? Add your label to Figure 1. What two phase transitions occur along this boundary line?

8. What phases are present in equilibrium at the triple point?

9. Along the vapor pressure curve, both the liquid and gas phases are present. If you look at a sample in a glass tube, you can see both these phases, just like looking at water in a bottle. You see the liquid at the bottom, and the water vapor (gas) is present in the space above it. As you increase the temperature, the vapor pressure increases. As you raise the temperature, what do you expect to see at the critical point?

Application

Phase diagrams for water and carbon dioxide are given below. Use these phase diagrams to answer the following questions about the properties of water and carbon dioxide. Note that the pressure scale for water is in torr, and the pressure scale for carbon dioxide is in atmospheres.

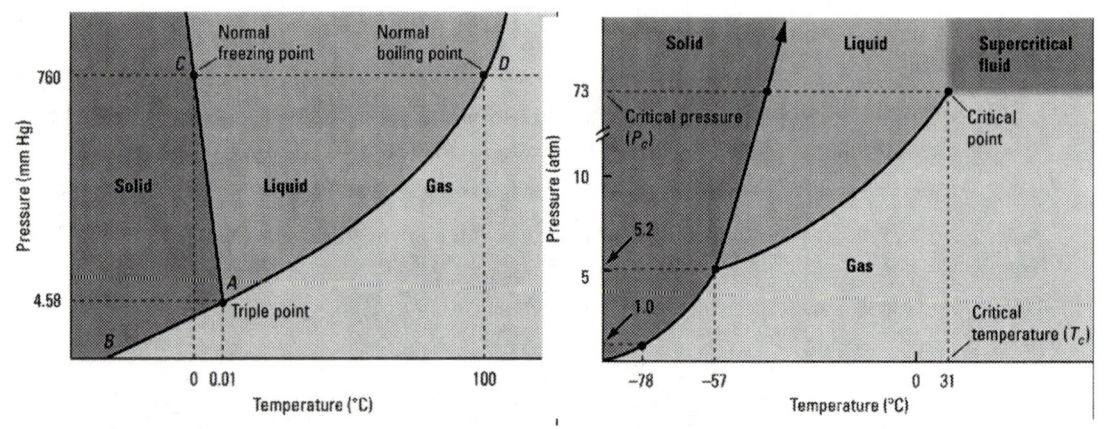

Fig. 5. Phase diagrams for water, H_2O (left), and carbon dioxide, CO_2 (right).

1. Use the phase diagram to estimate the vapor pressure of water at 50 °C.

2. Will ice sublime at some pressure and temperature? If not, why not? If so, below what temperature and pressure will ice sublime?

3. Label the triple point for carbon dioxide. What does this point represent?

4. Does the melting point of water increase or decrease with increasing pressure? Explain why in terms of the density of the liquid and solid.

5. Does the melting point of carbon dioxide increase or decrease with increasing pressure? Explain why in terms of the density of the liquid and solid.

6. (a) What is the lowest pressure and temperature at which liquid carbon dioxide is stable?

   (b) Identify the stable phase of carbon dioxide at temperatures below this point.

   (c) Identify the stable phase of carbon dioxide at pressures below this point.

Research

Explain why, in terms of noncovalent interactions, the triple point of carbon dioxide occurs at a higher pressure and a lower temperature than the triple point of water.

Got It!

1. How many phases are in equilibrium along a boundary line?

2. How can you identify the normal boiling point of a substance on a phase diagram if it isn't labeled?

3. How can you identify the normal melting point of a substance on a phase diagram if it isn't labeled?

4. How can you identify the critical point of a substance on a phase diagram if it isn't labeled?

5. Sketch a heating curve obtained for carbon dioxide at a pressure of 10 atm. Label each phase and the phase transitions on your heating curve.

General Chemistry: Guided Explorations

12-1. Fuels: How good is ethanol as a fuel?

Gasoline is mostly a mixture of hydrocarbons that are derived from petroleum. The 1990 amendments to the Clean Air Act of 1970 require the use of oxygenated gasoline to reduce carbon monoxide emissions when the gasoline burns. Gasoline is oxygenated by adding compounds that contain oxygen. Currently ethanol, CH_3CH_2OH, is being used for this purpose, and 5% to 10% ethanol by volume is added to the gasoline.

Even higher amounts of ethanol and also butanol, $C_4H_{10}O$, are being promoted as fuels to reduce our reliance on petroleum that is imported from other countries and to reduce the rate at which carbon dioxide is added to the atmosphere. A cost-conscious consumer wants the most "bang for the buck" and should know the amount of energy that is obtained from each gallon and dollar of fuel.

What do you think?

Isooctane, C_8H_{18}, is a good model for gasoline. Without doing any calculations, identify what you would need to consider in order to determine whether a gallon of ethanol produces more or less energy than a gallon of isooctane when burned?

Exploration

In this activity you determine the cost effectiveness of four fuels: ethanol, butanol, gasoline, and kerosene. Ethanol, butanol, and gasoline are used for automobile fuels. Kerosene with various additives is used in jet engines, and mixtures similar to kerosene are used in diesel engines and also for heating homes.

Gasoline is mostly a mixture of C_4 through C_{12} hydrocarbons, and its properties closely duplicate those of iso-octane, which is why gasoline has an octane rating. In this activity isooctane is used as the model for gasoline.

Kerosene is mostly a mixture of C_{11} through C_{15} hydrocarbons and is the principal ingredient in jet engine fuels. Diesel oil and fuel oil for heating are similar to kerosene in composition but have different additives, which produce a higher density and viscosity. In this activity dodecane, $C_{12}H_{26}$, is used as the model for kerosene, diesel oil, and fuel oil.

Three criteria are used to explore the merits of these four fuels: their fuel value (kJ/g), their energy density (kJ/mL), and their economic value (kJ/$). The *fuel value* is used to determine the energy obtained from a given mass, the *energy density* is used to determine the energy obtained from a given volume, and the *economic value* is used to determine the energy obtained for a given amount of money.

An inexpensive fuel is important in any application, and a high fuel value and energy density are important in applications where minimizing the mass and space devoted to fuel storage is important. Both racing cars and airplanes benefit from having fuels with a low mass (high fuel value) and a long range (high energy density).

Data that you need for this exploration are given in Table I. Standard enthalpies of combustion, $\Delta H°_{combustion}$, are used as a measure of the energy that can be obtained from these fuels.

Table I. Properties of Fuels

Fuel	Formula	Molar Mass (g/mol)	Density (g/mL)	$\Delta H°_{combustion}$ (kJ/mol)	Cost/gal ($/gal est.)
Ethanol	C_2H_5OH	46.07	0.789	-1,368	3.00
Butanol	C_4H_9OH	74.12	0.810	-2,670	3.00
Gasoline	C_8H_{18}*	114.22	0.668	-5,461	3.00
Kerosene	$C_{12}H_{26}$*	170.34	0.749	-7,900	3.00

*Iso-octane and dodecane) are used to model gasoline and kerosene, respectively.

1. Describe how you can determine the fuel value (kJ/g) for each fuel from the data in Table I.

2. Calculate the fuel value of each of the fuels and enter your results in Table II. The results for the case of ethanol are given in the table so you can check that your method of calculation is correct.

3. Describe how you can determine the energy density (kJ/g) for each of the fuels in Table I.

4. Calculate the energy density of each of the fuels and enter your results in Table II.

5. Describe how you can determine the economic value (kJ/$) for each of the fuels in Table I.

6. Calculate the economic value of each of the fuels and enter your results in Table II. Note: 1 gal = 3.785 L

Table II. Figures of Merit for Different Fuels

Fuel	Formula	Fuel Value (kJ/g)	Energy Density (kJ/mL)	Economic Value (kJ/$)
Ethanol	C_2H_5OH	29.7	23.4	29,560
Butanol	C_4H_9OH			
Gasoline	C_8H_{18}*			
Kerosene	$C_{12}H_{26}$*			

*Iso-ocatane and dodecane are used to model gasoline and kerosene, respectively.

7. Which fuel in Table II has the highest fuel value?

8. Which fuel in Table II has the highest energy density?

9. Which fuel in Table II has the highest economic value?

10. From the "most bang for the buck" perspective, which are the two best fuels in Table II?

11. From the information in Table II, identify two reasons why jet airplanes might use kerosene rather than gasoline.

Got It!

1. Use the information in Table II to explain why cars with diesel engines generally get better mileage than those with gasoline engines.

2. If a car gets 25 miles/gallon of gasoline, predict using the information in Table II how far an equivalent car could go using a gallon of diesel fuel.

3. If gasoline sells for $3.00 a gallon, what should the price of a gallon of ethanol be to have the same economic value?

4. A popular mixture of ethanol and gasoline is called E85. It consists of 85% ethanol and 15% gasoline by volume. If gasoline sells for $3.00 a gallon, what should the price of a gallon of E85 be to have the same economic value?

12-2. Organic Functional Groups: What determines the properties of complicated organic molecules?

A *functional group* is a combination of atoms bonded together in a characteristic pattern. The functional groups in an organic molecule are responsible for the chemical reactions that involve the molecule, its biological and medicinal activity, and the properties of the compound (e.g. solubility, vapor pressure, melting and boiling points),. For example, *verapamil*, which is shown below, is used to treat irregular heartbeats and high blood pressure. It has the following functional groups: 2 aromatic rings, 4 ethers, a nitrile, an amine, and alkane linkages.

For each functional group that you encounter in this activity, you need to remember its name and its characteristic bonding pattern. In a line drawing, like that show for verapamil, each line represents a bond, and a carbon atom is implicit wherever a line ends or lines come together. Carbon atoms always have 4 bonds, so bonds to hydrogen atoms are implicit wherever 4 bonds to a carbon atom are not shown.

Verapamil
$C_{27}H_{38}O_4N_2$

What do you know about functional groups already?

1. After reading the material above, identify five functional groups that you know by name and label them in the diagram of verapamil.

Exploration: Some Common Functional Groups

Complete Tables I and II by entering the distinguishing characteristics where they are missing.

Table I. Some Common Functional Groups

Structure	Name	Example	Distinguishing Characteristics
C—C	alkane	Propane, $CH_3CH_2CH_3$, a fuel	Carbon – carbon single bonds.
C=C	alkene	Ethylene, CH_2CH_2, used to make polyethylene	
C≡C	alkyne	Acetylene, CHCH, used in welding torches	
(hexagon ring)	aromatic ring	Benzene, C_6H_6, a carcinogen	A hexagon of six carbons with alternating double and single bonds made equivalent by resonance.
C—O—H	alcohol	Ethanol, CH_3CH_2OH, ingredient in beverages and gasoline	
C—X	halide	Carbon tetrachloride, CCl_4, X = any halogen used in dry cleaning.	
—N— (with lone pair)	amine	Ephedrine, a decongestant	
C—O—C	ether	Diethylether, $(CH_3CH_2)O$, an anesthetic	
N≡C	nitrile	Acetonitrile, CH_3CN, an important solvent	
C=O	carbonyl	Part of other groups shown in Table II.	

Table II. Carbonyl Containing Functional Groups

Structure	Name	Example	Distinguishing Characteristics
C–C(=O)–H	aldehyde	Formaldehyde, CH_2O, preservative for biological specimens	a carbonyl group with a carbon and hydrogen attached
C–C(=O)–C	ketone	Acetone, $(CH_3)_2O$, a product of fatty acid metabolism	
C–C(=O)–O–H	carboxyl	Acetic acid CH_3CH_2COOH, an ingredient in vinegar	a carbonyl group with an OH and carbon attached
C–C(=O)–O–X	acid halide. X = any halogen	Acetyl chloride, CH_3COCl, A very reactive compound used in synthesis	
C–C(=O)–O–C	ester	Ethyl acetate, nail polish remover	
C–C(=O)–N	amide	Nylon 6, a polymer used in fabrics, cords, and ropes	A carbonyl group with an amine group attached

Got It!

1. Para-aminobenzoic acid, PABA, is used in sun screen lotions. What are the three functional groups in this molecule?

109

2. What would you call this ingredient that is found in rhubarb: a carboxylic acid, a ketone, an alcohol, an ester, or an ether? Explain.

3. What are three functional groups present in aspirin?

4. What is different about these two steroids?

androstenedione   testosterone

5. Which of the following would <u>not</u> be a good hypothesis? Both of these molecules are narcotic pain relievers because they both have
   A) amine groups.
   B) similar structures.
   C) oxygen atoms in key positions.
   D) ester groups.

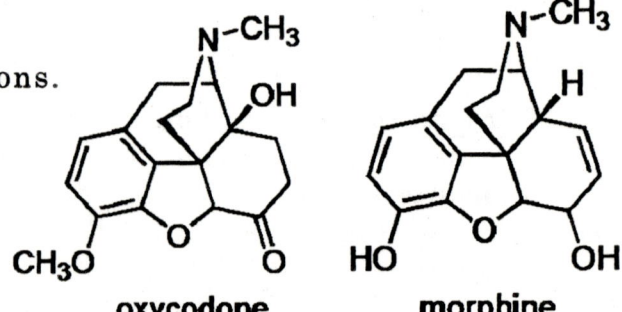

oxycodone   morphine

13-1. Rates of Chemical Reactions: How quickly do reactants turn into products?

The *rate* of any process is the change in some quantity over some period of time. For example, if you walk 1 mile over a period of 20 minutes, the rate at which your position changes is 1 mile/20 min = 0.05 mile/min, which also is called your speed. Usually such speeds are expressed in miles per hour, in these units your walking speed is (0.05 mile/min)(60 min/hr) = 3 mile/hr.

What do you think?

1. Identify 3 examples of something that you have observed changing over a period of time.

2. For each of the examples that you identified in #1 above, estimate the rate for that process.

Exploration

Peroxyacetylnitrate (PAN) is a powerful respiratory and eye irritant found in smog. It is produced from pollutants (unburned hydrocarbons, NO, and NO_2) in the exhaust of power plants and internal combustion engines.

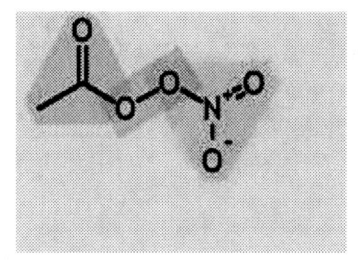

In an undergraduate research project, a team of students measured the rate of decomposition of PAN. Their results are given in Table I. The initial concentration of PAN was 50.0×10^{-6} M, and Table I shows that concentration decreased with time.

The concentration was monitored over a period of 91 minutes. The beginning of the experiment is designated as t = 0 (t = time).

Table I. Decomposition of Peroxyacylnitrate (PAN).

time (min)	conc (10^{-6} M)	<rate>	<rate>/conc
0	50.0		
1	48.3	1.740	0.0348
10	35.1		
11	33.9	1.221	0.0348
20	24.6		
21	23.8		
30	17.3		
31	16.7		
40	12.1		
41	11.7		
50	8.5		
51	8.2		
60	6.0		
61	5.8		
70	4.2		
71	4.0		
80	2.9		
81	2.8		
90	2.1		
91	2.0		

The students began to analyze the data by calculating the average rate of the decomposition reaction over a 1 minute interval after every 10 minutes by using Equation (1). The results of these calculations are in the third column of the table.

$$Rate = -\frac{\Delta[PAN]}{\Delta t} = -\frac{[PAN]_{t_2} - [PAN]_{t_1}}{t_2 - t_1} \qquad (1)$$

They also divided the average rate by the concentration at the beginning of each 1 minute interval. Those numbers are in the fourth column of the table.

1. The students ran out of time to complete their analysis. Please help them by providing the missing entries in the third and fourth columns of Table I. Enter the values that you calculate in the unshaded rectangles in the table.

2. Show how the concentration of PAN changes with time by making a graph in the space below. Plot the concentration of PAN on the y-axis and time on the x-axis. Be sure to label the axes, provide a scale for each axis, and include a title for the graph.

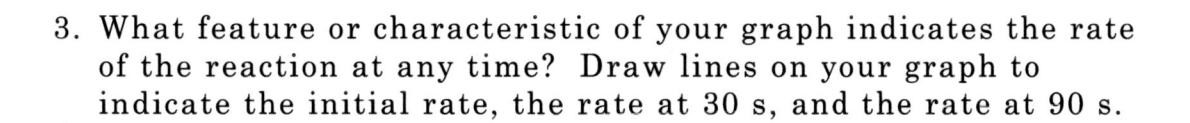

3. What feature or characteristic of your graph indicates the rate of the reaction at any time? Draw lines on your graph to indicate the initial rate, the rate at 30 s, and the rate at 90 s.

113

4. Does the rate of the reaction increase, decrease, or remain the same as the concentration of PAN decreases? How can you tell from your graph?

5. How long did it take for half the PAN to decompose? This time is designated as the half-lifetime or the *half-life* of the decomposition reaction. Draw a vertical line on your graph to mark the half-life time.

6. What happens to the ratio of the reaction rate to the concentration of PAN as the reaction progresses? This information is given in the fourth column of Table I.

7. In view of your answers to *Exploration* questions 1 – 6, which equation below, where k = a constant, describes the data in Table I? Explain or justify your choice.

a) Rate = k

b) Rate = k [PAN]

c) Rate = k [PAN]^2

Information

A rate law for a reaction tells how the rate of the reaction depends on the concentrations of the chemical species involved in the reaction. Rate laws have the form given by Equation (2). Equation (2) says that the rate of the reaction equals some constant multiplied by the product of concentrations [A], [B], and [C] of the reactants with exponents x, y, and z, respectively.

$$\text{Rate} = k[A]^x[B]^y[C]^z \qquad (2)$$

An exponent in the rate law gives the order of the reaction with respect to that chemical species. For example, if x and z equal 1 and y = 2, then the reaction is first order with respect to A and C and second order with respect to B. The overall order of the reaction is the sum of the exponents in the rate law, which in this example would be x + y + z = 1 + 2 + 1 = 4.

The choices in *Exploration* question 7 correspond to the rate laws for a *zero-order* reaction, a *first-order* or unimolecular reaction, and a *second-order* or bimolecular reaction. The rate of a zero-order reaction does not depend on the concentration of the reactants, the rate of a first-reaction depends linearly on the concentration, and the rate of a second-order reaction depends on the square of the concentration.

The order of a reaction often differs from the stoichiometric coefficients in the balanced reaction equation. The stoichiometric coefficients relate the numbers of reactant and product molecules involved in the reaction. The order of the reaction depends on the reaction mechanism, not on the number of molecules involved in the overall reaction.

The rate of a reaction always is expressed as a positive number and can be written in terms of the change in concentrations of any of the reactants or products. For example, for the reaction $2H_2 + O_2 \rightarrow 2H_2O$, the rate could be written in any of the following ways.

$$\begin{aligned} \text{a)} \quad & Rate = -\frac{1}{2}\frac{\Delta[H_2]}{\Delta t} \\ \text{b)} \quad & Rate = -\frac{\Delta[O_2]}{\Delta t} \\ \text{c)} \quad & Rate = \frac{1}{2}\frac{\Delta[H_2O]}{\Delta t} \end{aligned} \quad (3)$$

In (3a) and (3b) minus signs are needed to make the rate a positive number because the concentration is decreasing, so the change in concentration is negative. In (3a) and (3c) the change in concentration needs to be multiplied by ½ because those concentrations are changing twice as fast as the oxygen concentration and indicated by the stoichiometric coefficients in the balanced reaction equation.

Application

1. Is the decomposition of PAN a zero, first, or second order reaction? Explain.

2. Using the data in Table I, calculate the value and determine the units of the rate constant, k, for the decomposition of PAN.

3. Using the relationship $t_{1/2} = \ln(2)/k$, calculate the half-life, $t_{1/2}$, of PAN from the value you calculated for k in *Application* question 2. Is this calculated value consistent with your estimate of the half-life in *Exploration* question 5?

Information

The graph you made in the *Exploration* section has the form of an exponential function. This exponential function gives the concentration of PAN as a function of time.

$$[PAN(t)] = [PAN(t=0)]\, e^{-kt} \qquad (4)$$

Equation (4) is called the *integrated rate law* for a first order reaction. It is called the integrated rate law because its derivative gives the first order rate law, which you discovered in response to *Exploration* question 7.

$$Rate = -\frac{\Delta[PAN(t)]}{\Delta t} = k[PAN(t)] \qquad (5)$$

The integrated rate law, Equation (4), can be put in logarithmic form by taking the natural logarithm of both sides.

$$\ln\left([PAN(t)]\right) = -kt + [PAN(t=0)] \qquad (6)$$

Got It!

1. In view of Equation 6, what would you need to plot to produce a graph from the data in columns 1 and 2 of Table I so all the points lie on a straight line? Note that the equation for a straight line is y = ax + b.

2. Nickel is purified for use in making steel alloys by heating $Ni(CO)_4$ in a vacuum to decompose it in a unimolecular reaction: $Ni(CO)_4 \rightarrow Ni + 4CO$. At 66.0°C, the rate constant for this reaction is 1.873 s^{-1}.
(a) Write the rate law for this reaction.

(b) Calculate the time it would take for half the nickel tetracarbonyl to decompose.

(c) Calculate the time it would take for 90% of the nickel tetracarbonyl to decompose.

13-2. Reaction Mechanisms: What determines the rate law for a reaction?

Many chemical reactions do not occur in one step; rather, they involve a series of steps. The steps correspond to a single molecule dissociating or rearranging in some way or to two molecules colliding and reacting with each other. Reaction mechanisms rarely involve the simultaneous collision of three or more molecules because it is extremely unlikely for three or more molecules to arrive at the same point in space at the same time. Such an event is similar to three cars arriving from different directions and colliding at an intersection, collisions of two cars at an intersection occur much more often.

Each step in the overall reaction is called an *elementary reaction*, and a reaction equation can be written for it. The set of reaction equations for all the elementary reactions is called the *reaction mechanism*.

A rate law can be predicted for any reaction mechanism, and the predicted rate law can be compared with the one observed experimentally. If the predicted and experimental rate laws disagree, then the proposed reaction mechanism is not the correct one. If they agree, one can only say that the mechanism is consistent with the experimental rate law. Agreement does not prove that the proposed mechanism is the correct one because different mechanisms can predict the same rate law.

What do you think?

Based on the previous activity, *Rates of Chemical Reactions*,

1. What is the rate law for a unimolecular reaction, e.g. A → B?

2. What is the rate law for a bimolecular reaction, e.g. A + B → C?

Exploration 1: Predicting a Rate Law

Nitric oxide, NO, is present in the exhaust of power generating plants and internal combustion engines. It reacts with the oxygen in air to produce nitrogen dioxide, NO_2, which causes smog to have a brownish color.

119

$$2NO(g) + O_2(g) \rightarrow 2NO_2(g)$$

Two mechanisms that have been proposed for this reaction are given below.

Mechanism 1
Step 1: $NO + O_2 \rightleftharpoons NO_3$ (fast, equilibrium established)
Step 2: $NO_3 + NO \rightarrow 2NO_2$ (slow)

Mechanism 2
Step 1: $NO + O_2 \rightarrow NO_2 + O$ (slow)
Step 2: $NO + O \rightarrow NO_2$ (fast)

1.1. Any valid mechanism must consist of a series of unimolecular or bimolecular elementary reaction steps that give the overall reaction equation when added together.
(a) Add the 2 steps for Mechanism 1 together. Do they give the overall reaction equation?

(b) Add the 2 steps for Mechanism 2 together. Do they give the overall reaction equation?

1.2. The rate of the overall reaction is limited by and equal to the rate of the slowest step in the mechanism.
(a) What is the rate law for the slow step in Mechanism 1?

(b) What is the rate law for the slow step in Mechanism 2?

1.3. To compare with experimental measurements, the rate law for a reaction must be expressed in terms of the concentrations of the reactants and not include the concentrations of intermediates. Do the rate laws that you have written in response to *Exploration* question 2 include only the reactants NO_2 and O_2?

1.4. To eliminate an intermediate from the rate law, assume that the fast reaction is reversible and that the concentration of the intermediate reaches a steady state or equilibrium value. In the steady state, the intermediate is destroyed as fast as it is formed. The following equation describes this steady state condition for Step 1 in Mechanism 1.

$$k_1[NO][O_2] = k_{-1}[NO_3] \qquad (1)$$

Rearranging this equation gives

$$[NO_3] = (k_1/k_{-1})[NO][O_2] \qquad (2)$$

(a) Using Equation (2), rewrite your the rate law that you predicted in *Exploration* question (1.2a) for Mechanism 1 in terms of only the concentrations of NO and O_2.

(b) Write the rate laws that you predict for Mechanisms 1 and 2 in the table below.

Table I. Rate Laws Predicted for the Oxidation of Nitric Oxide

Mechanism 1	Rate =
Mechanism 2	Rate =

Exploration 2: Determining the Rate Law Experimentally

The relationship between the concentration of reactants and the rate of the reaction is given by the rate law for the reaction. One way to determine this relationship experimentally is to measure the initial rate of the reaction when different concentrations of reactants are used. Such data collected for the oxidation of nitric oxide are given in Table II.

The rate law for the oxidation of nitric oxide has the following form: Rate = k $[NO]^x[O_2]^y$. The objective is to use the data in Table II to determine the exponents x and y in the rate law.

Table II. Initial Reaction Rates for the Oxidation of Nitric Oxide

Experiment	[NO] 10^{-2} M	[O$_2$] 10^{-2} M	Rate 10^{-3} M s^{-1}
A	1	1	7
B	2	1	28
C	1	2	14

2.1. According to the data in Table II, when the concentration of nitric oxide was kept constant and the concentration of oxygen increased by a factor of 2, by what factor did the initial rate of the reaction increase?

2.2. In view of your answer to Question 2.1, what is the value of the exponent y in the rate law, Rate = k [NO]^x[O$_2$]^y?

2.3. According to the data in Table II, when the concentration of nitric oxide increased by a factor of 2 and the concentration of oxygen was kept constant, by what factor did the initial rate of the reaction increase?

2.4. In view of your answer to Question 2.3, what is the value of the exponent x in the rate law, Rate = k [NO]^x[O$_2$]^y?

Got It!

1. Write the experimentally determined rate law for the oxidation of nitric oxide.

2. Identify which reaction mechanism, 1 or 2, is consistent with the experimental rate law.

3. Calculate a value for the rate constant for the oxidation of nitric oxide from the data in Table II.

14-1. Equilibrium Constant and Reaction Quotient: How can the direction of a reaction be identified?

When chemicals are mixed together, they generally react to produce products. The concentrations of the reactants decrease and the concentrations of the products increase until dynamic equilibrium is established. When dynamic equilibrium is established, the concentrations of reactants and products no longer change because products are being converted back to reactants at the same rate as reactants are being converted into products.

This equilibrium condition is defined by an *equilibrium constant* for every reaction. For the following general reaction at equilibrium,

$$aA + bB \rightleftharpoons cC + dD \qquad (1)$$

the *equilibrium constant expression* is written as,

$$K_c = \frac{[C]_{eq}^c [D]_{eq}^d}{[A]_{eq}^a [B]_{eq}^b}. \qquad (2)$$

The equilibrium constant is a ratio of the molar concentrations of the products divided the molar concentrations of the reactants with each concentration raised to the power given by the corresponding stoichiometric coefficient in the balanced reaction equation. The numerical value of this ratio is called the equilibrium constant.

The *reaction quotient* has the same mathematical form as the equilibrium constant expression except it involves the actual concentrations in a reaction mixture instead of the equilibrium concentrations.

$$Q = \frac{[C]^c [D]^d}{[A]^a [B]^b} \qquad (3)$$

What do you think?

The concentration of products in a chemical reaction stops increasing when
(a) all the reactants have been converted into products.
(b) the concentrations of reactants and products are equal.
(c) the products are converted back into reactants at the same rate as reactants are being converted to products.

Exploration

1. What items go into the numerator of the equilibrium constant expression?

2. What items go into the denominator of the equilibrium constant expression?

3. What determines the exponent for each concentration in the equilibrium constant expression?

4. What are three similarities in the expressions for the equilibrium constant and the reaction quotient?

5. What do you see as the most important difference between the expressions for the equilibrium constant and the reaction quotient?

6. Would you expect a reaction with a large value for the equilibrium constant to be reactant favored or product favored? Explain.

7. If Q is larger than K (Q > K) for a given reaction mixture, would you expect the reaction to proceed in the forward direction to produce more products or in the reverse direction to produce more reactants? Explain.

8. For a given reaction mixture, when will a value calculated for Q equal the value for K found in a table of equilibrium constants?

Application

1. In an automobile engine, nitrogen and oxygen combine to form nitric oxide. Write the equilibrium constant expression, K_c, for this reaction.

$$N_2(g) + O_2(g) \rightleftharpoons 2\ NO(g)$$

2. Calcium ions react with carbonate ions in water to form a solid deposit of calcium carbonate around water faucets. Write the equilibrium constant expression for this reaction. Note that the concentration of a pure solid (mol/L) does not change during a reaction; consequently, the concentration of the solid is not included in the equilibrium constant expression.

$$Ca^{2+}(aq) + CO_3^{2-}(aq) \rightleftharpoons CaCO_3(s)$$

3. Acetic acid is an ingredient in vinegar. It reacts with water to produce hydronium ions and acetate ions. Write the equilibrium constant expression for this reaction. Note that the concentration of water is very large (55.5 mol/L) and does not change significantly as a result of the reaction, so the concentration of water or any other pure liquid is not included in the equilibrium constant expression.

$$CH_3COOH(aq) + H_2O(l) \rightleftharpoons H_3O^+(aq) + CH_3COO^-(aq)$$

Got It!

1. Explain how the value of the equilibrium constant can help you decide whether or not products predominate over reactants when equilibrium has been achieved for any reaction.

2. Given the concentrations of reactants and products, explain how the values of the equilibrium constant and the reaction quotient can help you identify the direction of the reaction, i.e. whether the reaction will go forward to produce more products or reverse to produce more reactants.

3. The equilibrium constant for the reaction of hydrogen, H_2, with iodine, I_2, to form hydrogen iodide, HI, has a value of 50.0 at 745 K.
   (a) What is the value of the reaction quotient for a mixture of 0.5 mol H_2, 0.5 mol I_2, and 5.0 mol HI in a 2 L container?

   (b) Will more HI be formed or will HI decompose as the reaction mixture approaches equilibrium? Explain.

14-2. Calculating Equilibrium Concentrations: How can the concentrations of reactants and products at equilibrium be determined?

In order to make coal a cleaner and more transportable fuel, methods are being developed to produce combustible gases from it. In one of these methods carbon monoxide produced by burning coal with a limited amount of oxygen is reacted with steam to produce hydrogen.

$$CO(g) + H_2O(g) \rightleftharpoons CO_2(g) + H_2(g)$$

Researchers in this area need to know how much carbon monoxide and water can be converted into hydrogen under various conditions. For example, if 1 mol each of carbon monoxide and steam were placed in a 20.0 L container at 420°C, what would the concentrations of the four gases be when equilibrium is reached. The equilibrium constant for this reaction is 10 at 420°C.

Such equilibrium problems are easily solved if you use a systematic procedure. One such procedure involves constructing a table of amounts, as shown by Table I.

To determine the concentrations at equilibrium, an unknown amount, x, is assumed to react to reach equilibrium (row 3 in Table I).

Table I. Table of Amounts for the Reaction of CO and H$_2$O

CO(g)	H$_2$O(g)	$\rightleftharpoons$	CO$_2$(g)	H$_2$(g)	Write the balanced reaction equation.
0.05	0.05		0	0	Initial concentrations (mol/L)
-x	-x		+x	+x	Change caused by the reaction (mol/L)
0.05-x	0.05-x		+x	+x	Concentrations at equilibrium (mol/L)

The concentrations at equilibrium (row 4 in Table I) then are substituted into the equilibrium constant expression and the resulting equation is solved to find a value for x, the amount that reacts, as shown below.

$$K_c = \frac{[CO_2][H_2]}{[CO][H_2O]}$$

$$10 = \frac{x^2}{(0.05-x)^2}$$

Taking the square root of both sides produces

$$3.16 = \frac{x}{0.05-x}$$

Rearranging and solving for x produces

$3.16(0.05 - x) = x$

$x = 0.158 / 45.16$

$x = 0.0380$

The equilibrium concentrations therefore are 0.012 M for both CO and H₂O, and 0.038 M for both CO₂ and H₂.

Exploration

1. Why is writing the balanced reaction equation an important part of the procedure for solving equilibrium problems?

2. How would the third and fourth lines in Table I be changed if the stoichiometric coefficients in the balanced reaction equation were different than 1?

3. Why is the last line in the Table I an especially important one?

4. What insight about solving equilibrium problems has your team gained by examining the above example and procedure?

Information

Generally substituting the equilibrium concentrations into the equilibrium constant expression will produce a quadratic equation for x. A quadratic equation has the following form, where a, b, and c are constants.

$$ax^2 + bx + c = 0$$

A quadratic equation has 2 possible solutions, which are given by the following equation. While both solutions are mathematically valid, generally only one of the solutions makes physical sense and is consistent with the situation.

$$x = \frac{-b \pm \sqrt{b^2 - 4ac}}{2a}$$

In *Application* 2 below, K_c is small, which means that the amount that reacts, x, is small. You then can solve the equation by assuming that x is negligible, and [NOCl] – x = [NOCl]. You generally can make this assumption if the equilibrium constant is less than 10^{-3}, and if it turns out that the amount that reacts divided by the initial concentration is less than 0.05.

Application

1. Carbon dioxide reacts with carbon to give carbon monoxide.

    $$C(s) + CO_2(g) \leftrightarrows 2\, CO(g)$$

    At 700°C, a 2.00 L flask contains an equilibrium mixture of 0.10 mol CO(g), 0.20 mol CO_2(g), and 0.40 mol C(s). Calculate the equilibrium constant, K_c, for this reaction at 700°C.

2. At a certain temperature the equilibrium constant for the decomposition of NOCl is $K_c = 2.0 \times 10^{-5}$. Calculate the equilibrium concentrations of NO and Cl_2 when 1.00 mol of NOCl is placed in a 1.00 L container.

    $$2NOCl(g) \leftrightarrows 2\, NO(g) + Cl_2(g)$$

Got It

1. Why was it not necessary to solve a quadratic equation to determine the equilibrium concentrations of CO, H_2O, CO_2, and H_2 in the example at the beginning of the activity?

2. Why is it often not necessary to solve a quadratic equation in equilibrium problems when the equilibrium constant is small?

3. What amount of hydrogen iodide, HI, will be produced when 1 mol of iodine, I_2, and 3 mol of hydrogen, H_2, are mixed in a 10.0 L container at 745 K? $K_c = 50.0$ at 745°C for the following reaction.

$$H_2(g) + I_2(g) \leftrightarrows 2\ HI(g)$$

15-1 Solubility: Where do those vitamins go?

What do you think?

Dietary supplements, such as vitamins, are a multi-billion dollar industry. Over 40% of the adult population takes dietary supplements with vitamins C and E being among the most popular. What do you think you should know about vitamins before deciding whether or not to take them as a dietary supplement?

Information

Vitamins are small molecules that are essential for our health. Most of the vitamins that we need are obtained from food and are not synthesized in our bodies. Vitamins are divided into two categories: those that are water soluble (hydrophilic) and those that are not water soluble (hydrophobic). The hydrophilic vitamins dissolve in the blood. The hydrophobic vitamins do not dissolve in the blood directly but are transported by blood lipoproteins and are absorbed by fatty membranes and tissues.

Some people advocate taking mega doses of vitamins to improve health, but large doses of the hydrophobic vitamins can cause harmful effects because they are stored with fat in cells and dissolve in fatty tissues, including cell membranes. Consequently, a significant concentration is maintained in the bloodstream, and the concentrations in the tissues build up over time. The hydrophilic vitamins, on the other hand, are not soluble in the fatty tissues; they remain dissolved in the blood and are eliminated quite rapidly through the kidneys. Consequently, they do not build up high concentrations in the body.

Vitamin C is an antioxidant and a coenzyme. It is known that it prevents scurvy. As an antioxidant, it neutralizes reactive molecules in cells (free radicals), reduces damage to the cells, and thereby prevents cancer. As a coenzyme, it contributes to the enzymatic synthesis of other essential molecules, such as collagen, which is the main connective protein in joints.

Vitamin E actually is a group of 8 isomers with structures similar to the one shown in Figure 1. A number of studies indicate that vitamin E, which functions as a powerful antioxidant, may prevent heart disease, cancer, cataracts, Alzheimer's disease, and age-related macular degeneration.

The structures of vitamin C and vitamin E are shown below.

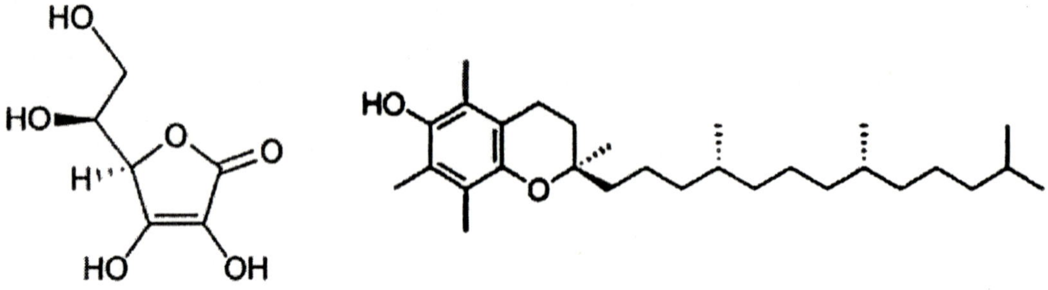

Fig. 1.   Vitamin C (ascorbic acid)   Vitamin E

Exploration

1. Using the structures in Figure 1, identify the types of intermolecular interactions that are possible for vitamin C and for vitamin E. (Consider London dispersion, dipole-dipole, hydrogen bonding, and ionic).

    Vitamin C:

    Vitamin E:

2. Which type of intermolecular interaction do you expect to be most important for vitamin C? Explain.

3. Which type of intermolecular interaction do you expect to be most important for vitamin E? Explain.

4. Considering your answers to *Exploration* questions 2 and 3 above, which vitamin do you expect to be hydrophilic and which one do you expect to be hydrophobic? Explain.

5. Considering your answers to *Exploration* questions 2 and 3 above and the structures shown in Figure 2 below, which vitamin, C or E, do you expect to be absorbed better by fatty tissue and scavenge free radicals in phospholipids cell membranes? Explain.

Fig.2.   triglyceride fat   phospholipid cell membrane

6. Based on the discussion in the *Information* section and your identification of the vitamins C and E as either hydrophilic or hydrophobic, which one do you think will remain in the bloodstream longer? Explain.

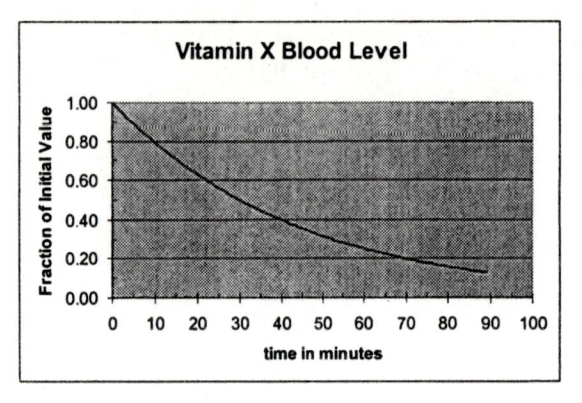

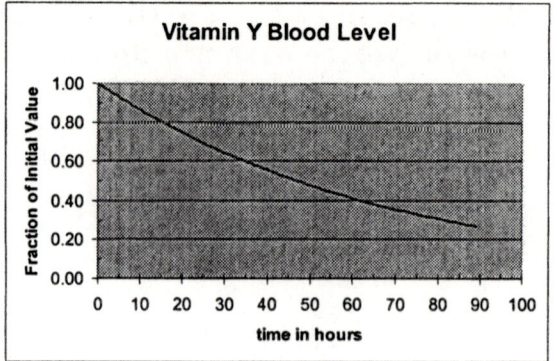

Vitamin X Blood Level (time in minutes, 0–100, fraction decreases from 1.00 to ~0.15)

Vitamin Y Blood Level (time in hours, 0–100, fraction decreases from 1.00 to ~0.27)

Fig. 3. Graphs showing the decrease in concentration of two vitamins, labeled X and Y, in the blood as a function of time after the initial dose. Note the difference in the x-axis scale.

7. Replace X and Y in the titles of the graphs in Figure 3 with the name of the corresponding vitamin, C and E, as appropriate. Explain your choice.

8. From the graphs in Fig. 3, determine the half-life of each vitamin in the bloodstream.

9. How long will it take for each vitamin concentration to drop to 10% of its initial value in the bloodstream?

10. How often do you think you need to take each vitamin to keep the concentrations above 50% of the initial concentration? Explain.

11. Draw concentration vs time graphs showing how repeated doses of each vitamin would keep the concentrtion between 50% and 150% of the initial peak concentration.

12. Based your analysis, what is your recommendation for how often you should take Vitamin C and E to maintain a consistent level of each vitamin in your bloodstream? Explain.

Got It!

1. Predict which one of the following vitamins would be retained in the bloodstream longer. Explain.

   Vitamin B3, niacin   Vitamin D

2. Amoxicillin and azythromicin are common antibiotics. Amoxicillin usually is taken every 6 hours for 10 days, while azythromicin is taken once a day for 5 days. Based on these regimens, what can you predict about their half-lives and molecular structures. Explain.

Reflection on Learning and Performance

1. Identify 5 things that you or your team learned in this activity.

2. Identify 5 skills that you or your team needed to exercise in completing this activity.

3. What could you have done differently that would have helped you complete this activity more efficiently and effectively?

Acknowledgement

This activity was adapted from a draft titled, *Vitamin C and E: How long do I have to stay here?* that was produced by Renee Cole, Linda Hobart, and Fred Yost at the POGIL-IC workshop, Litchfield, South Carolina, January, 2007.

15-2 Colligative Properties:

What do you think?

If you add salt to water does the water
(a) boil at a higher or lower temperature?

(b) freeze at a higher or lower temperature?

Information

Colligative properties are properties of dilute solutions that depend only on the number of nonvolatile solute particles and not on the type of particles present. They include a lowering of the vapor pressure, elevation of the boiling point, depression of the freezing point, and osmotic pressure. All these effects are caused by adding a nonvolatile solute to a solvent.

Since in a given amount of solution, some solvent molecules have been replaced by nonvolatile solute particles, fewer solvent molecules will be able to reach the surface and escape. Consequently the vapor pressure of the solution will be smaller.

The boiling point is the temperature at which the vapor pressure equals 1 atm. Since the solute has lowered the vapor pressure, then the boiling point of the solution necessarily will be higher because a higher temperature is needed to produce a vapor pressure of 1 atm.

Depression of the freezing point and osmotic pressure also can be traced to a decrease in vapor pressure caused by the nonvolatile solute.

Table I. Colligative Properties of Solutions

Property	Equation	Symbols
Vapor Pressure	$P = XP^0$	P = vapor pressure of the solvent over the solution P^0 = vapor pressure of the pure solvent X = mole fraction of the solvent in the solution
Boiling Point Elevation	$\Delta T_b = K_b m i$	ΔT_b = change in boiling point K_b = solvent boiling point elevation constant m = molal concentration of the solute i = van't Hoff factor = the number of moles of particles produced by 1 mole of solute
Freezing Point Depression	$\Delta T_f = -K_f m i$	ΔT_f = change in freezing point K_f = solvent freezing point depression constant m = molal concentration of the solute i = van't Hoff factor = the number of moles of particles produced by 1 mole of solute
Osmotic Pressure	$\Pi = cRTi$	Π = osmotic pressure c = molar concentration of the solute R = gas constant T = temperature i = van't Hoff factor = the number of moles of particles produced by 1 mole of solute

Exploration

1. What is the distinguishing characteristic of a colligative property?

2. What are four colligative properties of dilute solutions?

3. What are three things that these colligative properties have in common that can be identified from Table I?

4. How can one use the formula in Table I for the osmotic pressure to determine the molar mass of a polymer by dissolving it in some solvent?

Application

1. People living at high altitudes where the pressure is less than 1 atm, e.g. Denver, often add salt to water to increase the boiling temperature when cooking. Is this a significant effect? Check by determining the boiling point of 1.0 L of water when 1 tablespoon of NaCl is added to it.

2. People living in cold climates, e.g. Minnesota, often throw some salt on icy sidewalks to melt the ice. If $CaCl_2$ is used, what is the concentration of the resulting solution if the ice melts at $-10°C$?

3. Order the following aqueous solutions in order of increasing osmotic pressure:
0.10 M KCl, 0.10 M sucrose, 0.05 M K_2SO_4, and 0.10 M $CaCl_2$.

4. Calculate the average molar mass of polyethylene when 4.40 g of the polymer is dissolved in benzene to produce 200.0 mL of solution and the osmotic pressure is found to be 7.60 torr at 25°C.

5. Sulfur exists in many forms with the general molecular formula S_n. If 0.48 g of sulfur is added to 200 g of carbon tetrachloride, CCl_4, and freezing point is depressed by 0.28°C, what is the molar mass and molecular formula of the sulfur?

Got It!

1. Two beakers, one (labeled A) containing 100.0 mL of a 1.0 M salt solution (NaCl), and the other (labeled B) containing 100.0 mL of pure water are placed inside a larger container, which then is sealed.
   (a) What do you think happens to the liquid levels in the two beakers as time elapses? Explain your reasoning?

   (b) Sketch a graph showing how the concentration of the solution in Beaker A changes with time.

2. (a) In what ways is the experiment in *Got It* question 1 similar to an osmosis experiment where two compartments of a container are separated by a membrane that only allows solvent molecules to pass. On compartment is filled with pure solvent, and the other compartment is filled with a salt solution.

   (b) What happens to the liquid levels in those two compartments as time elapses?

16-1. Acid Ionization Constants: How strong is that acid?

Svante Arrhenius first defined an *acid* as a substance that produces hydronium ions (H_3O^+) when added to water, and a *base* as a substance that produces hydroxide ions (OH^-) when added to water.

Later Johannes Brønsted in Denmark and T. Martin Lowry in England independently proposed to give acids and bases more general definitions. According to the Brønsted-Lowry concept, an *acid* is a substance than donates a hydrogen ion or proton (H^+) to another substance, and a *base* is a substance that accepts a proton from another substance.

What do you think?

Identify some acids that you know are found in food products, e.g. in vinegar, citrus fruits, grapes, apples, and milk.

Information

In an aqueous solution, Brønsted-Lowry acids react with water and ionize. For example,

$$HF(aq) + H_2O(l) \rightleftharpoons H_3O^+(aq) + F^-(aq). \tag{1}$$

In this reaction, water acts as a base and accepts a proton from a hydrofluoric acid molecule.

Such ionization reactions are described by an equilibrium constant, which is called the *acid ionization constant*, K_a. The acid ionization constant corresponding to the ionization of hydrofluoric acid in Equation (1) is defined as

$$K_a = \frac{[H_3O^+][F^-]}{[HF]} \tag{2}$$

Values for some acid ionization constants are given in Table I.

Table I also contains pK_a values. pK_a is defined in Equation (3) and is used because it is quicker to write and enter into a calculator than the K_a value.

$$pK_a = -\log(K_a). \tag{3}$$

Table I. Acid ionization constants at 25°C

Acid	Formula	K_a	pK_a
Hydrochloric acid	HCl	1.0×10^4.	-4.0
Sulfuric acid	H_2SO_4	1.0×10^3	-3.0
Nitric acid	HNO_3	2.0×10^1	
Phosphoric acid	H_3PO_4	7.5×10^{-3}	2.1
Hydrofluoric acid	HF	7.2×10^{-4}	
Nitrous acid	HNO_2	4.5×10^{-5}	4.3
Acetic acid	CH_3COOH		4.7
Carbonic acid	H_2CO_3	4.2×10^{-7}	6.4
Dihydrogen phosphate	$H_2PO_3^-$	6.2×10^{-8}	7.2
Ammonium ion	NH_4^+	5.6×10^{-10}	9.3
Ethanol	CH_3CH_2OH	1.0×10^{-16}	16.0
Methane	CH_4	1.0×10^{-48}	48.0

Exploration

1. What is the definition of a Brønsted-Lowry acid?

2. What is the definition of a Brønsted-Lowry base?

3. Considering both the forward and reverse reactions in Equation (1), what are the two Brønsted-Lowry acids in this reaction equation?

4. Considering both the forward and reverse reactions in Equation (1), what are the two Brønsted-Lowry bases in this reaction equation?

5. What determined the order from top to bottom in which the acids are written in Table I?

6. Some Ka and pKa values are missing in Table I. Use the one that is given to calculate, using Equation (3), the one that is missing, and complete the table.

7. For a 1.0 M solution of hydrofluoric acid, [HF] = 0.973 M, [H$_3$O$^+$] = 0.0264 M, and [F$^-$] = 0.0264 M. Show whether or not these concentrations are consistent with the acid ionization constant for hydrofluoric acid in Table I.

8. For a 1.0 M solution of hydrochloric acid, [HCl] = 0.0001 M, [H$_3$O$^+$] = 0.9999 M, and [Cl$^-$] = 0.9999 M. Show whether or not these concentrations are consistent with the acid ionization constant for hydrochloric acid in Table I.

9. To produce the concentrations in *Exploration* question 7, 0.0264 moles of the 1 mole of HF ionized. What percent of the HF molecules ionized?

10. To produce the concentrations in *Exploration* question 8, 0.9999 moles of the 1 mole of HCl ionized. What percent of the HCl molecules ionized?

11. Hydrofluoric acid is called a *weak acid* and hydrochloric acid is called a *strong acid*. Based on your results for *Exploration* questions 9 and 10, why do you think hydrofluoric acid is called a weak acid, and why do you thing hydrochloric acid is called a strong acid?

Got It!

1. Based on your answer to *Exploration* question 11, write a definition for a *weak acid*.

2. Based on your answer to *Exploration* question 11, write a definition for a *strong acid*.

3. Using your definition of a strong acid, list all the strong acids that are given in Table I.

4. Using your definition of a weak acid, list all the weak acids that are given in Table I.

5. (a) Complete Table II by identifying each of the acids as strong or weak and calculating the K_a value.
   (b) List the acids in Table II in order of decreasing strength.

Table II. Some acids found in foods.

Acid	Strong/Weak	Formula	pK_a	K_a
Lactic acid – found in milk products			3.88	
Oxalic acid – found in rhubarb			1.23	
Malic acid – found in sour or tart foods like green apples			3.40	

16-2. Calculations Involving Acid Ionization Constants: How can the pH of an acid solution be determined from the K_a value?

An acid ionization constant, K_a, is the equilibrium constant for an acid ionization reaction. It provides information about the concentrations of the reactants and products at equilibrium.

An important product in an acid ionization reaction is the hydronium ion, and the hydronium ion concentration usually is expressed as a pH value using Equation (1).

$$pH = -\log([H_3O^+]) \qquad (1)$$

What do you think?

In a previous activity (*14-2 Calculating Equilibrium Concentrations*) you learned how to connect the equilibrium constant to the equilibrium concentrations. The strategy involved constructing a *Table of Amounts*. Summarize below the steps in this strategy.

Exploration 1: Determining the pH of a Strong Acid Solution

What is the pH of a solution prepared by adding 0.25 mol of HCl to 125 mL of water? In the *Table of Amounts* on the following page:

(a) Write the balanced reaction equation.

$$(HCl + H_2O \rightleftharpoons H_3O^+ + Cl^-) \qquad (2)$$

(b) Write the number of moles of each species present before the reaction occurs below that species. (0.25 moles of HCl, note that water doesn't appear in K_a so water can be ignored)

(c) Write the number of moles of each species that reacts or is produced. (HCl is a strong acid so all of it, 0.25 moles, reacts, and 0.25 moles of hydronium and chloride ions are produced.)

(d) For reactants, subtract the moles that react from the initial moles to find the equilibrium moles. For products, add the moles that are produced to the initial moles to obtain the equilibrium moles.

(e) Divide the equilibrium moles by the volume of the solution to obtain the equilibrium concentrations.

Table I. Table of Amounts for Strong Acid Ionization

a) Reaction equation	
b) Initial moles	
c) Change in moles	
c) Equilibrium moles	
e) Equilibrium conc	

Exploration 2: Determining pK$_a$ from pH.

Lactic acid, $C_3H_6O_3$, is found in a number of food products. If a 0.25 M solution has a pH of 2.23, what is the pK$_a$ for lactic acid?

In completing the *Table of Amounts*, initially there were 0.25 moles of lactic acid in 1.0 L of solution. The moles of hydronium ion, which can be determined from the pH, equals the number of moles of lactic acid that ionized and also the moles of lactate produced. Use the equilibrium concentrations that you find to calculate K$_a$ and pK$_a$.

Table II. Table of Amounts for pKa of lactic acid

a) Reaction equation	
b) Initial moles	
c) Change in moles	
c) Equilibrium moles	
e) Equilibrium conc	

Exploration 3: Determining pH from K_a

Carbon dioxide dissolves in water to form carbonic acid, H_2CO_3. What is the pH of natural rainwater that is saturated with carbon dioxide? The pK_a for carbonic acid is 6.38, and the concentration of carbonic acid in a saturated solution is 1.09×10^{-5} M.

The initial amount of carbonic acid is 1.09×10^{-5} moles in 1.0 L of solution. The amount that reacts to form hydronium ion is what you need to find, call this amount x. The equilibrium amount of carbonic acid is $(1.09 \times 10^{-5}) - x$ moles, but the x can be neglected because K_a is small, and consequently the amount that ionizes, x, will be small compared to the amount present initially.

In situations where this approximation is not valid, the expression for the equilibrium constant produces a quadratic equation in x that can be solved using the quadratic formula.

Table III. Table of Amounts for the pH of rainwater.

a) Reaction equation	
b) Initial moles	
c) Change in moles	
c) Equilibrium moles	
e) Equilibrium conc	

Got It!

1. Why is writing the balanced reaction equation an important part of the *Table of Amounts* strategy especially when some of the stoichiometric coefficients differ from 1?

2. In the above problem dealing with the ionization of lactic acid, why are the hydronium ion and lactate concentrations equal and determined by the pH?

3. In the above problem dealing with the pH of rainwater, why is the amount of carbonic acid present at equilibrium taken to be the amount of carbonic acid present initially despite the fact that some of the carbonic acid reacts and ionizes?

4. Do you prefer a different strategy for solving these kinds of problems? In what way is your strategy different?

5. What insight about solving acid equilibrium problems did you or your team gain from this activity so far?

6. Ephedrine hydrochloride, which provides the active ingredient in some decongestants, is a salt like ammonium chloride $NH_4^+Cl^-$. We use the formula EH^+Cl^- for this salt. The cation EH^+ has a $pK_a = 9.96$. The pH of blood (7.4) determines the hydronium ion concentration in the ephedrine acid ionization equilibrium.

$$EH^+ + H_2O \leftrightarrows H_3O^+ + E \qquad (3)$$

(a) Calculate a value for the ratio $[E]/[EH^+]$ in the bloodstream.

(b) Which form, the acid EH^+ or its conjugate base E, is predominantly present in the bloodstream? Explain

17-1. Buffers: How can your blood pH be kept near the critical value of 7.4?

Many reactions and structures in biological systems, especially those involving proteins and enzymes, require that a biomolecule be either in the acid form with a proton or in the base form without the proton in order to be active and functional. The ambient pH determines which form is present. If the pH is too low or too high, then the equilibrium will be shifted in the wrong direction, and life processes will cease! Consequently it is essential for biological systems to have mechanisms for controlling and maintaining the correct pH. One of these mechanisms is the use of buffers.

The pH of an aqueous solution of a weak acid and the salt of its conjugate base can be calculated from the equilibrium constant expression. For example, consider 1.0 L of solution containing 0.01 mole of NaF and 0.01 mole of HF, which has $K_a = 7.2 \times 10^{-4}$. The equilibrium constant expression for the ionization reaction

$$HF(aq) + H_2O(l) \rightleftharpoons H_3O^+(aq) + F^-(aq) \tag{1}$$

is

$$K_a = \frac{[H_3O^+][F^-]}{[HF]}. \tag{2}$$

Since K_a and the HF and F^- concentrations are known, Equation (2) can be solved for the hydronium ion concentration and then the pH calculated.

Equation (2) also can be manipulated to produce a new equation that is a bit easier to use. Take minus log of both sides and use the equality $\log(ab) = \log(a) + \log(b)$ to obtain equation (3).

$$-\log(K_a) = -\log[H_3O^+] - \log\left(\frac{[F^-]}{[HF]}\right). \tag{3}$$

Then replace the minus log quantities with pK_a and pH and rearrange to obtain Equation (4). This equation is known as the Henderson-Hasselbalch equation.

$$pH = pK_a + \log\left(\frac{[F^-]}{[HF]}\right) \tag{4}$$

Exploration

1. The pH of neutral water at 25°C is 7.0 because the concentrations of hydronium ion and hydroxide ion are both equal to 1.0×10^{-7} M. What is the pH of the solution if 0.010 mole of HCl is added to 1.0 L of water?

2. What is the pH of the solution if 0.01 mole of sodium hydroxide is added to 1.0 L of water?

3. Use Equation (4) to calculate the pH of 1.0 L of solution containing 0.50 mole HF and 0.50 mole NaF. Note that although some of the HF ionizes, the amount that does is negligible because K_a is so small.

4. Calculate the pH of the solution in Exploration question 3 after 0.01 mole of hydrochloric acid, HCl, has been added to it. Construct a *Table of Amounts* to determine the new equilibrium concentrations of HF and F⁻. Note that the 0.01 mole of HCl reacts with 0.01 mole of F- to produce an additional 0.01 mole of HF.

5. Calculate the pH of the solution in Exploration question 3 after 0.01 mole of a strong base, sodium hydroxide NaOH, has been added to it. Note that the base converts 0.01 mole of HF to F⁻.

6. Complete the table below to summarize the results of your calculations in Exploration questions 1 – 5.

Table I. Effect of a strong acid and a strong base on pH

Solution	Amount	pH
Pure water	1.0 L	7.0
HCl added	0.01 mole added	
NaOH added	0.01 mole added	
Solution 0.50 M in both HF and NaF	1.0 L	
HCl added	0.01 mole added	
NaOH added	0.01 mole added	

7. In view of your calculated results that are summarized in Table I, what can you say about the relative effects of a strong acid and a strong base on the pH of the HF/NaF solution compared to the effect on the pH of pure water?

8. Why do you think the addition of a strong acid or a strong base to the HF/NaF solution had a much smaller effect on the pH compared to the effect with pure water?

9. Suppose you wanted to make up another solution that resisted pH change just like the HF/NaF solution did, but had a pH value closer to 7. What would you do to make such a solution?

Information

A *buffer* or a *buffer solution* is a chemical system that resists changes in pH when an acid or a base is added to it or when it is diluted. A buffer solution is composed of a weak acid and its conjugate base. The pH of a buffer solution is determined by the acid's pK$_a$ value and the ratio of base/acid concentrations as given by the Henderson-Hasselbalch equation.

Got It!

1. According to Equation (4), what will determine the pH of a buffer be if the weak acid and its conjugate base are present in equal amounts?

2. Calculate the pH of a solution 0.0150 M in acetic acid and 0.300 M in sodium acetate.

3. The pH of blood is buffered by several mechanisms to maintain a value close to 7.4. Calculate the ratio HPO_4^{2-} / $H_2PO_4^-$ that is needed to buffer at pH = 7.4. K$_a$ = 6.2 10^{-8} for dihydrogen phosphate.

4. A buffer consists of 0.30 M propanoic acid (K$_a$ = 1.4 x 10^{-5}) and 0.30 M sodium propanoate.
   (a) What is the pH of this buffer?

   (b) What is the pH after 1.0 mL of 0.10 M HCl is added to 10 mL of the buffer?

   (c) What is the pH after 10.0 mL of 0.10 M HCl is added to 10.0 mL of the buffer?

   (d) Compare your answers to parts (b) and (c). What happens to the effectiveness of the buffer as more and more acid or base is added?

17-2. Acid – Base Titrations: How much acetic acid is there in an exotic wine vinegar?

In a titration, a substance in a solution with a known concentration reacts with a second substance in another solution. The stoichiometry of the reaction is known, and the objective is to determine the amount or concentration of the second substance. The progress of the reaction is monitored by some indicator that reveals when the equivalence point has been reached. The *equivalence point* is the point in the titration when the two substances have reacted completely so that neither is present in excess.

For this activity, you might imagine that you are working in the analytical laboratory of *Nature's Own Natural Foods, Inc.* Your task is to analyze the titration of acetic acid with sodium hydroxide to determine whether or not the titration can be used reliably to measure the concentration of acetic acid in your company's exotic vinegars. To demonstrate the validity of the titration procedure, you need to calculate the pH at various points in the titration to demonstrate that existing acid-base theory is consistent with experimental titration data.

Exploration

To explore and validate the titration procedure, consider what happens when 25.0 mL of 0.100 M acetic acid, CH_3COOH, reacts with increasing amounts of 0.050 M sodium hydroxide, NaOH. Acetic acid is a weak acid with a pK_a of 4.74.

1. What is the initial pH of the acetic acid solution? The initial pH is the pH of the solution before any sodium hydroxide is added. Use the acid ionization constant expression to calculate the hydronium ion concentration and the pH of the 0.100 M acetic acid solution.

2. What is the pH of the acetic acid solution after 10.0 mL of the sodium hydroxide solution has been added to it? Sodium hydroxide reacts with acetic acid to produce acetate ions. Determine how much acetic acid remains and how much acetate has been produced, then use the Henderson-Hasselbalch equation to calculate the pH.

3. What is the pH after 25.0 mL of the sodium hydroxide solution has been added? Determine how much acetic acid remains and how much acetate has been produced, then use the Henderson-Hasselbalch equation to calculate the pH.

4. What is the pH after 45.0 mL of the sodium hydroxide solution - has been added?

5. What is the pH after 50.0 mL of the sodium hydroxide solution has been added? You should see that this is the equivalence point. There were 25.0 mL x 0.100 M = 2.50 mmol of acetic acid and now 50.0 mL x 0.05 M = 2.50 mmol of sodium hydroxide have been added. Essentially all of the acetic acid has been converted into acetate ions (Ac⁻). Now, the equilibrium constant expression for the hydrolysis of acetate, K_b, can be used to calculate the hydroxide ion concentration, which then can be converted to the pOH and the pH.
$$Ac^-(aq) + H_2O(l) \rightleftharpoons HAc(aq) + OH^-(aq) \qquad pK_b = 9.26$$

6. What is the pH after 55.0 mL of the sodium hydroxide solution has been added? The sodium hydroxide in 50.0 mL of solution reacted with all of the acetic acid, so now there is 5.00 mL of excess sodium hydroxide or 0.25 mmol in 80.0 mL of solution. The concentration of excess sodium hydroxide is the hydroxide ion concentration, which can be converted to the pOH and the pH.

7. What is the pH after 75.0 mL of the sodium hydroxide ion solution has been added? Determine the amount of sodium hydroxide in excess, and divide by the volume of the final solution to get the hydroxide ion concentration.

8. Summarize the results of you calculations from Exploration questions 1 – 7 in the following table. For major species, identify acetic acid, acetate, or both as the major species with the highest concentrations that determine the pH.

Table I. Calculated data for the titration of 25.0 mL of 0.100 M acetic acid with 0.05 M sodium hydroxide.

Situation	Major Species	$[H_3O^+]$	pH
Before adding NaOH solution			
10.0 mL NaOH			
25.0 mL NaOH			
45.0 mL NaOH			
50.0 mL NaOH			
55.0 mL NaOH			
75.0 mL NaOH			

9. Draw a graph below of your data in Table I. Plot pH on the y-axis and the volume of the NaOH solution added on the x-axis. A high quality graph has a title and axes with labels and scales. Points are plotted with small circles around them to make them easy to see, and a smooth line is drawn through the data points to show how the trend continues between the points.

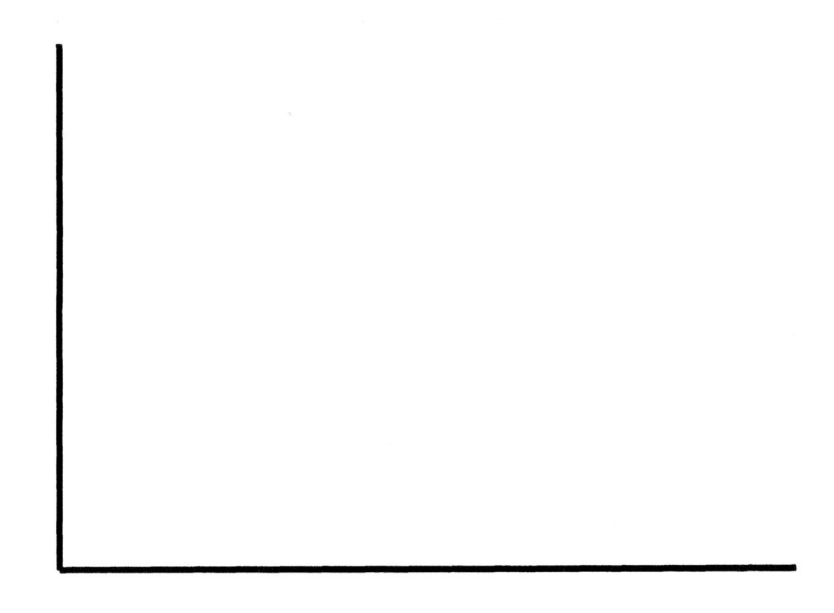

Got It!

1. *Nature's Own Natural Foods, Inc.*'s premier product is champagne vinegar. Your lab technician titrated 50.0 mL of this specialty item with 0.500 M sodium hydroxide and found the equivalence point at 84.0 mL.

   (a) What is the molar concentration of acetic acid in NONF's champagne vinegar?

   (b) What is the mass percent of acetic acid in this product, which has a density of 1.010 g/mL? The molar mass of acetic acid is 60.05 g/mol.

   (c) Is the amount of acetic acid in NONF's champagne vinegar consistent with the label, which claims 5% (by weight)? Explain.

2. At what point in the titration of a weak acid with a strong base is the pH equal to the pK$_a$ of the acid? Explain.

3. Where in the titration of a weak acid with a strong base is a buffer solution produced?

4. Why does the pH change so abruptly near the equivalence point?

18-1. Entropy: Why do some things occur spontaneously while others do not?

Every process has a preferred direction, which is called the spontaneous direction. For example, snow melts on a warm day, iron rusts when in contact with water and oxygen, and water runs downhill. A *spontaneous process* is an action that occurs on its own accord unless restrained in some way, and work must be done or energy expended in order to reverse a spontaneous process. For example, water runs downhill unless restrained by a dam, and a pump is needed to move water uphill.

Exploration – 1

Table I. Characteristics of Spontaneous and Non-spontaneous Processes

Process	Initial State	Final State	ΔH	Spontaneous
Methane burning $CH_4(g)+2O_2 \rightarrow CO_2(g)+2H_2O(g)$	molecules in gas phase	molecules in gas phase	exothermic	yes
Water decomposing $2H_2O(g) \rightarrow 2H_2(g)+O_2(g)$	molecules in gas phase	molecules in gas phase	endothermic	no
Salt dissolving in water $NaCl(s) \rightarrow Na^+(aq)+Cl^-(aq)$	ordered crystal	solvated ions dispersed in solution	endothermic	yes
Ice melting $H_2O(s) \rightarrow H_2O(l)$	ordered crystal	disordered liquid	endothermic	yes, above 0°C
Water evaporating $H_2O(l) \rightarrow H_2O(g)$	disordered liquid	disordered gas, expanded volume	endothermic	yes

1.1. Using the characteristics of spontaneous and non-spontaneous processes in Table I and others that you might think of, identify what spontaneous processes have in common that distinguishes them from non-spontaneous processes.

What do you think?

What causes some processes to occur spontaneously?

Information

A spontaneous process can be either endothermic or exothermic, but whenever a process is spontaneous, the process causes the energy to disperse or spread out over more states. One also can say that more states are made accessible by the process.

Now apply this idea of energy dispersal and increase in the number of accessible states to the examples in Table 1. The combustion of methane is spontaneous because it is exothermic. Energy released in the reaction is dispersed over states of the surroundings. Salt dissolving in water is endothermic, but there are more positions (states) accessible to the ions in the solution than in the solid so this reaction also is spontaneous. Similarly ice melting and water evaporating are spontaneous because the number of possible positions (states) of the molecules in space increases.

The idea of the number of accessible states and the dispersal of energy making processes spontaneous is quantified in the concept of *entropy*. Entropy is a measure of the number of accessible states and the dispersal of energy. Ludwig Boltzmann and Rudolf Clausius, Austrian and German physicists, made these connections in the late 19th century.

Boltzmann defined entropy by the equation

$$S = k \ln(\Omega) \quad (1)$$

where S is the entropy, k is a fundamental constant now known as the Boltzmann constant (1.3807×10^{-23} J/K), and Ω is the number of accessible microstates of the system. A *microstate* is a state of an atom or molecule. For example, the quantum numbers $n, l, m_l,$ and m_s specify the microstates of the hydrogen atom.

Boltzmann's idea was that processes evolve spontaneously to produce the most probable situation, the one with the largest number of possible or accessible microstates. From this perspective, the total entropy for any spontaneous process must increase. The change in the total entropy, ΔS_{total}, is the change in the entropy of the system that is changing spontaneously, ΔS_{system}, plus the change in the entropy of the surroundings, $\Delta S_{surroundings}$. Often the *total entropy* is called the *entropy of the universe*.

$$\Delta S_{universe} = \Delta S_{system} + \Delta S_{surroundings} \quad (2)$$

Through his studies of heat engines, Clausius discovered that entropy is related to the dispersal of energy. A heat engine is a machine that converts heat into work, e.g. an automobile engine or a steam engine. A diagram of the key features of a heat engine is shown in Figure 1. This diagram illustrates that some amount of energy, q_H, is transferred into a heat engine at a high temperature, T_H. Some portion of this energy is transformed into work, W, and the remainder, q_L, is transferred out of the heat engine at a low temperature, T_L. The engine then returns to its original state, as illustrated by the curved arrow in Figure 1, and the cycle of heat in, heat out, work done repeats. In order to optimize the performance of a heat engine, Clausius collected data on the energy transferred and work done under various conditions. Data similar to that collected by Clausius is given in Table II.

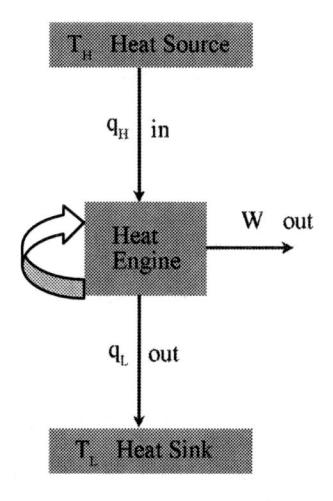

Fig. 1 Features of a heat engine.

Table II. Heat Engine Data, Rudolf Clausius, 1867

T_H	q_H	T_L	q_L	q_H/T_H	q_L/T_L
500 K	+10 kJ	250 K	-5 kJ	+20 J/K	-20 J/K

Exploration – 2

2.1. What are 4 patterns that you can find in the data in Table II?

2.2. Which pattern do you see as the most significant? Why?

Information

In examining the information in Table II, perhaps you noticed that T_L and q_L are both less than T_H and q_H, respectively; that the energy flow q changes sign at the lower temperature; that the temperature decreases by ½ and so does the amount of energy transferred; and that the sum of q/T for the two temperatures is 0.

Clausius too recognized something really amazing in his experiments! Like you, he noticed that the sum of q/T for the two temperatures is 0.

$$q_H/T_H + q_L/T_L = 0 \qquad (3)$$

Heat engines run in repetitive cycles, and after each cycle the engine returns to the same state. After each cycle the change in the quantity q/T is zero, which means that this quantity is a *state function* because it only depends on the state and not on what has happened.

Clausius called this quantity the *entropy*, gave it the symbol S, and proposed that the change in entropy for any reversible process at constant temperature is given by the energy flow divided by the Kelvin temperature.

$$\Delta S = q/T \qquad (4)$$

He chose the term *entropy* from a Greek word meaning *to put into transformation*. He liked this term because he was working on the transformation of heat into work and because it sounded similar to energy and enthalpy, which also originate from Greek words meaning *to put into work* and *to put into heat*, respectively.

Got It!

Water was put into a freezer and spontaneously froze at 0°C to produce 9.0 g of ice (0.5 mol). The temperature of the freezer was -15°C (258 K). The enthalpy of fusion of ice is 6.01 kJ/mol.

1. When the water froze, did energy flow from the water to the freezer or from the freezer to the water?

2. Use Equation (4) to calculate a value for the change in entropy of the system (water and ice). Note that q_{system} is negative if energy flowed into the surroundings and positive if energy flowed into the system.

3. Calculate a value for the change in entropy of the surroundings (the freezer).

4. Determine the change in entropy of the universe as a result of 9.0 g of water freezing in this freezer at -15°C.

5. Write a paragraph describing why ice forming in a freezer at −15 °C is spontaneous. In your description, identify the direction of energy flow, whether the number of accessible states of the system (the water and ice) increases or decreases, whether the number of accessible states of the surroundings (the freezer) increases or decreases, and whether the entropy change for the system, the surroundings, and the universe each were positive or negative. Include an explanation with each of your statements.

Application

1. What is the change in entropy of water when 36.0 g vaporizes at 100°C. The enthalpy of vaporization for water is 40.7 kJ/mol.

2. Which one in the following pairs will have the larger number of accessible states and hence the larger entropy. A system with a large number of possible states often looks like it is disordered so sometimes greater entropy is associated with more disorder.

   (a) solid argon vs liquid argon

   (b) liquid argon vs argon gas

   (c) argon gas at 100°C vs argon gas at 200°C

   (d) N_2 and O_2 in separate 1.0 L containers vs the same amount of N_2 and O_2 in a single 2.0 L container (Hint: If a valve were opened between the two 1.0 L containers, would the N_2 and O_2 mix spontaneously?)

   (e) 1 mol C(g) and 1 mol O_2(g) vs 1 mol CO_2(g)

   (f) 1.0 L of 1 M NaCl(aq) vs 2.0 L of 0.5 M NaCl(aq)

   (g) solid sugar in a cup of coffee vs the same amount of sugar dissolved in a cup of coffee

   (h) 1 mol He(g) vs 1 mol N_2(g) (Hint: Atoms have translational states available to them; molecules have vibrational and rotational states available in addition to translational states.)

   (i) C_2H_2(g) vs C_6H_6(g) (Hint: The number of vibrational states increases with the number of atoms in a molecule.)

18-2. Free Energy: How much work can be obtained from any process?

What do you think?

Can a process be spontaneous at one temperature but not at another? Explain or provide an example to support your answer.

Information

The entropy change of a chemical reaction, $\Delta S_{reaction}$, depends on the difference in the entropy of the products and reactants.

$$\Delta S_{reaction} = S_{products} - S_{reactants} \qquad (1)$$

The entropy of the surroundings, $\Delta S_{surroundings}$, also changes as a result of a chemical reaction because energy is transferred either into the surroundings from the system (an exothermic reaction) or into the system from the surroundings (an endothermic reaction).

A chemical reaction is spontaneous if it causes an increase in the entropy of the universe, $\Delta S_{universe}$, where

$$\Delta S_{universe} = \Delta S_{surroundings} + \Delta S_{reaction} \qquad (2)$$

An exothermic reaction ($\Delta H < 0$) transfers energy to the surroundings, thereby making more states accessible, and increasing the entropy. An endothermic reaction ($\Delta H > 0$) removes energy from the surroundings, thereby reducing the number of accessible states, and decreasing the entropy. The entropy change of the surroundings can be calculated from Clausius's definition $\Delta S = q/T$ by using Equation (3).

$$\Delta S_{surroundings} = - \Delta H_{reaction} / T \qquad (3)$$

Combining Equations (2) and (3) gives Equation (4)

$$\Delta S_{universe} = - \Delta H_{reaction} / T + \Delta S_{reaction} \qquad (4)$$

The right-hand side of Equation (4) depends only on properties of the reaction: the enthalpy and entropy changes and the temperature. Rather than call this quantity $\Delta S_{universe}$, J. Willard Gibbs defined a new quantity that we now call the Gibbs free energy, G. Gibbs decided to define the change in free energy as minus T times the change in the entropy of the universe,

$$\Delta G_{reaction} = - T \Delta S_{universe}. \qquad (5)$$

With this definition, Equation (4) becomes

$$\Delta G_{reaction} = \Delta H_{reaction} - T\,\Delta S_{reaction} \qquad (6)$$

Since each quantity in Equation (6) has the subscript *reaction*, it is not necessary to include the subscript

Although the situation for chemical reactions is considered explicitly in the above discussion, the conclusion can be applied to any process, and ΔG is called the free energy because it is the maximum amount of work that can be obtained from any process.

$$\Delta G = \Delta H - T\,\Delta S \qquad (7)$$

Exploration

1. What quantity gives the amount of heat that is produced or absorbed by a chemical reaction?

2. In Equation (3), why is it - $\Delta H_{reaction}$ and not just $\Delta H_{reaction}$?

3. Since ΔG for the reaction 2H₂O(g) → 2H₂(g) + O₂(g) is positive, which reaction is spontaneous: the formation of water from hydrogen and oxygen or the decomposition of water to produce hydrogen and oxygen? Explain.

Got It!

Some reactions are spontaneous at any temperature, others are not spontaneous at any temperature, some are spontaneous only at low temperatures, and others are spontaneous only at high temperatures. These four characteristics are a consequence of the four combinations of ΔH and ΔS that are possible. Complete the following table by identifying whether ΔG will be positive or negative and which of the above 4 characteristics result from the combination of ΔH and ΔS given for each row of the table.

Table I. Effects of ΔH, ΔS, and temperature on reaction spontaneity.

ΔH	ΔS	ΔG = ΔH −TΔS	Spontaneous always, never, at low T, or at high T?
> 0	> 0		
> 0	< 0		
< 0	> 0		
< 0	< 0		

19-1. Electrochemical Cells: Where does the energy come from to power your cell phone?

Batteries supply electrical current to power various devices that you use. On the street and in Radio Shack usually no distinction is made between an electrochemical cell and a battery, but actually a battery consists of several electrochemical cells connected together to produce a higher voltage. For example, a 12 volt automobile battery consists of six 2 volt cells connected together. An electrochemical cell also is called a voltaic cell or a galvanic cell.

What do you think?

Batteries power various devices that you use. What is the source of the electrical power from a battery?
(A) Electrons stored on a metal film inside the battery.
(B) Ions moving from one end of the battery to the other through the external circuit.
(C) Electrons produced by one chemical reaction are consumed in another reaction.
(D) Electrons produced by a chemical reaction are consumed in powering the device.

Exploration

1. Identify the components of a Zn/Cu electrochemical cell from Figure 1 and complete the table below.

Component	Anode Compartment	Cathode Compartment
Electrode metal (Cu or Zn)		
Electron potential (+ or -)		
Type of half reaction		
Write the half reaction		

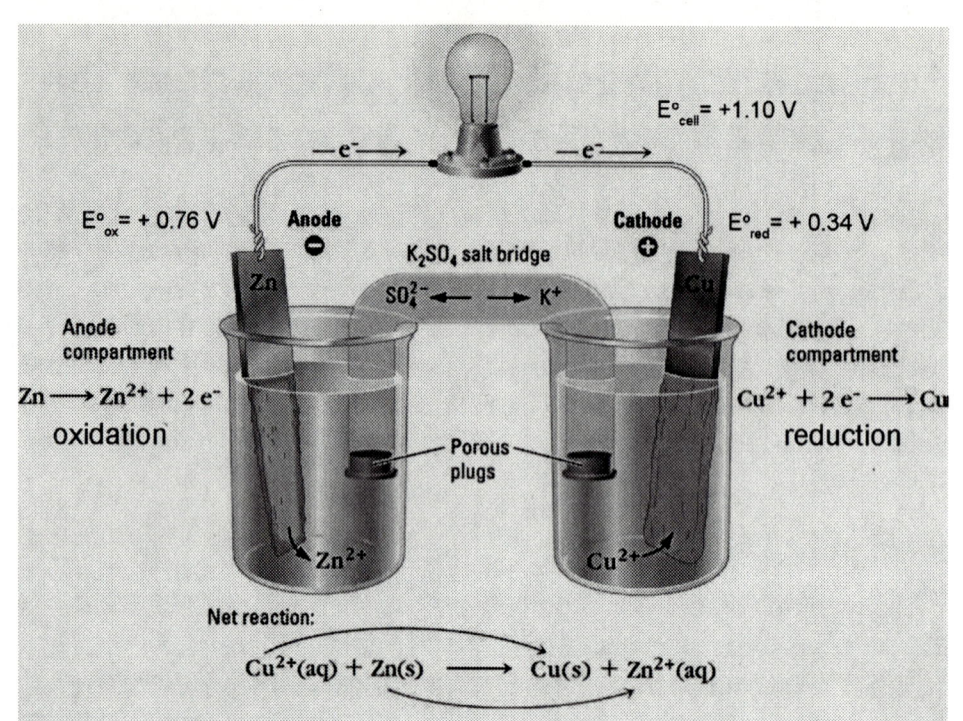

Fig. 1. An electrochemical cell being used to light a light bulb. This cell has Zn and Cu metal electrodes immersed in 1.0 M ZnSO₄ and CuSO₄ solutions, respectively.

2. Write the reaction that produces electrons to power the light bulb.

3. Write the reaction consumes the electrons.

4. Which way do the electrons flow in the external circuit, from the anode to the cathode or from the cathode to the anode?

5. Is the concentration of cations in the anode compartment increasing or decreasing?

6. In view of your answer to Exploration question (5), which way do positive and negative ions flow in the salt bridge? Explain why.

7. What is the difference between an oxidation reaction and a reduction reaction?

8. The voltage produced by an electrochemical cell is called the cell potential, E_{cell}. Examine Fig. 1, and identify how the standard cell potential, $E°_{cell}$, is determined from the standard half-cell potentials for the oxidation, $E°_{ox}$, and reduction, $E°_{red}$ reactions.

9. Read the information section below, and then write an equation that you can use to calculate a standard cell potential from the standard reduction half-cell potentials.

Information

The standard cell potential, $E°_{cell}$, is a measure of the force that attracts electrons to the positive electrode. The total driving force for the cell is determined by the sum of the driving forces for the oxidation reaction and for the reduction reactions. The cell potential therefore is determined by the half-cell potentials for these two reactions, $E°_{ox}$ and $E°_{red}$.

By convention, reference tables list standard reduction potentials and the corresponding reduction reactions. The half-reaction with the more positive the reduction potential has the greater force to attract electrons. For example,

Cu^{2+}(1M) + 2e⁻ → Cu(s) $E°_{red}$ = + 0.34 V
Zn^{2+}(1M) + 2e⁻ → Zn(s) $E°_{red}$ = − 0.76 V

Since the copper reaction has the more positive reduction potential, it runs as a reduction and the zinc reaction reverses and runs as an oxidation. When a half reaction is reversed, the value of the half-cell potential remains the same, but the sign reverses, so for a given half reaction, $E°_{ox}$ = − $E°_{red}$.

Got It!

A silver oxide battery, which is commonly used in watches, is based on the following half reactions.

$Ag_2O(s) + H_2O(l) + 2e^- \rightarrow 2Ag(s) + 2OH^-(aq)$ $E°_{red} = + 0.34$ V

$ZnO(s) + H_2O(l) + 2e^- \rightarrow Zn(s) + 2OH^-(aq)$ $E°_{red} = - 1.22$ V

1. Which reaction will run as a reduction and which one will run as an oxidation in the silver oxide battery?

2. What is the standard cell potential of the silver oxide battery?

3. Write the reactions occurring at the anode and at the cathode and label them as *oxidation* or *reduction*.
   anode reaction:

   cathode reaction:

4. Write the overall reaction for the silver oxide battery. You add the reduction reaction and the oxidation reaction together to get the overall reaction.

5. Identify the species being oxidized and the species being reduced in the overall reaction.

6. Identify the oxidizing agent and the reducing agent in the overall reaction.

7. Identify which of the following does not occur in the silver oxide battery.
   (a) Electrons flow from Zn(s) to $Ag_2O(s)$.
   (b) Silver metal is produced as the battery operates.
   (c) Corrosive hydroxide ion (OH-) is produced.
   (d) Ions move from one electrode to the other.

19-2. Electrolytic Cells: How can silver be deposited on nickel to make a beautiful friendship ring?

An electrochemical cell uses the energy derived from chemical reactions to produce electrical energy. An electrolytic cell, on the other hand, uses electrical energy to drive chemical reactions. Electrolysis is a very important industrial process. It is used to refine and purify many metals, like copper and aluminum, and it is used to plate thin coatings of one metal on top of another. An example of electrolysis, which is shown in Figure 1, is the decomposition of sodium chloride to produce chlorine gas and sodium metal.

What do you think?

Compare Figure 1 below with Figure 1 in Activity 19 on electrochemical cells. What do you see as the two most significant differences between an electrochemical and an electrolytic cell?

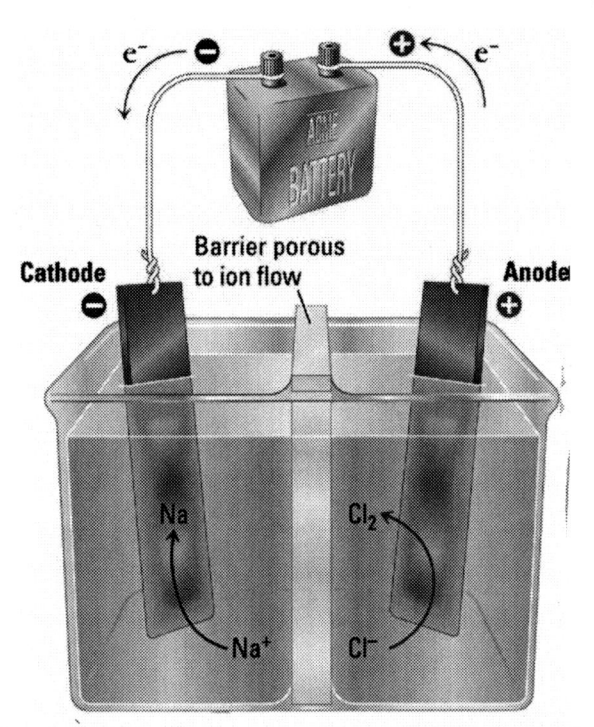

Fig. 1. Electrolysis of molten sodium chloride.

Exploration - 1

1.1. Write the half reaction occurring at the anode in Figure 1.

1.2. Write the half reaction occurring at the cathode in Figure 1.

1.3. Write the overall reaction for the electrolysis of sodium chloride.

1.4. What are the standard reduction potentials for Cl_2 and Na^+? Use your textbook or other resource to find these values.

1.5. Use your answer to Exploration question (5) to determine the minimum voltage that that is needed for the electrolysis of sodium chloride under standard conditions. As for an electrochemical cell, you add the standard potentials for the oxidation reaction and the reduction reaction together.

1.6. Why do electrochemical and electrolytic cells involving the same half reactions have standard cell potentials with the same magnitude but different signs?

Exploration – 2

The amount of electrical current that flows through an electrochemical cell and the change in the amount of reactants and products present are related. Exploration questions 2.1 and 2.2 will help you see how these quantities are related by determining the number of moles of sodium and chlorine that will be produced if an electrical current of 10 amperes flows through the electrolytic cell in Figure 1 for 10 hours.

2.1. Use the balanced reaction equation to determine the number of moles of sodium, Na, and the number of moles of chlorine, Cl_2, that are produced by the transfer of 1 mole of electrons.

2.2 Use the following information to determine the number of moles of electrons that are transferred by 10 A of current flowing for 10 hours.
   A = ampere = unit of electrical current
   C = coulomb = unit of electrical charge
      1 A = 1 C/s
   F = Faraday's constant = charge on 1 mole of electrons
      F = 96,485 C/mol
   charge transferred = current x time
   moles transferred = charge transferred / Faraday's constant

2.3. From your results for Exploration questions (2.1) and (2.2), how many grams of sodium metal and how many liters of chlorine gas as STP are produced by the electrolysis of sodium chloride using 10 A for 10 hours?

Got It!

From one perspective, hydrogen obtained from water is an ideal fuel. A photovoltaic cell can convert solar energy (sunlight) into electricity. The electricity can be used to electrolyze water to produce hydrogen and oxygen. When the hydrogen and oxygen recombine, water is produced and the energy is released for use. By these means, the solar energy can be stored, transported, and recovered. The whole process is self-sustaining and non-polluting.

The electrolysis of water involves the following two half reactions.

$$2H_3O^+(aq) + 2e^- \rightarrow H_2(g) + 2H_2O(l)$$

$$O_2(g) + 4H_3O^+(aq) + 4e^- \rightarrow 6H_2O(l)$$

1. Write the overall balanced reaction equation for the electrolysis of water.

2. Identify the species being oxidized and the species being reduced.

3. Determine the minimum voltage that would be required to electrolyze water under standard conditions using data in a table of Standard Reduction Potentials.

4. Determine the time it will take to produce 22.4 L of hydrogen gas, H_2, at STP using a current of 1 A.

20-1. Radioactivity: Are there different kinds of radioactivity?

Not all numbers of protons and neutrons in the nucleus of an atom are stable. The nuclei that are unstable spontaneously transform or decay into more stable nuclei by emitting particles or electromagnetic radiation. This process is called *radioactivity* and the unstable atoms are called *radioactive*. Some of these decay pathways are listed in Table I. They involve the emission of alpha, beta, and positron particles plus electromagnetic radiation called gamma rays. An alpha particle is the nucleus of a helium atom, which consists of 2 protons and 2 neutrons. A beta particle is an electron, and a positron is a particle with the same mass as an electron but with a charge of +1.602 x 10^{-19} C rather than the electron charge of – 1.602 x 10^{-19} C.

What do you think?

Radioactivity is both useful and dangerous. Identify 3 beneficial uses of radioactive substances and a reason why radioactivity can be dangerous.

Exploration - 1

Table I. Nuclear Decay Processes

Process	Symbol	ΔZ	ΔA	Example
Alpha particle emission	α			$^{234}_{92}U \rightarrow {}^{230}_{90}Th + \alpha$
Beta particle emission	β			$^{239}_{92}U \rightarrow {}^{239}_{93}Np + \beta$
Gamma ray emission	γ			$^{57}_{26}Fe^* \rightarrow {}^{57}_{26}Fe + \gamma$ (* = excited state)
Positron emission	$^{0}_{1}e$			$^{207}_{84}Po \rightarrow {}^{207}_{83}Bi + {}^{0}_{1}e$
Electron capture	$^{0}_{-1}e$			$^{7}_{4}Be + {}^{0}_{-1}e \rightarrow {}^{7}_{3}Li$

Exploration

1. Review atomic symbol notation, and identify what the left superscript and the left subscript tell you, for example in $^{234}_{92}U$.

2. The atomic symbol notation for a positron is $^{0}_{1}e$. Write the atomic symbol notation with a left superscript and a left subscript for an alpha particle, which is a helium nucleus, and for a beta particle, which is an electron.

3. Complete Table I by adding the entries for ΔZ and ΔA, where Z is the atomic number and A is the mass number

4. What is the relationship between the mass numbers of the reactants and the mass numbers of the products in a nuclear decay process?

5. What is the relationship between the atomic numbers of the reactants and the atomic numbers of the products in a nuclear decay process?

6. Table I contains examples where a proton appears to turn into a neutron and a neutron appears to turn into a proton. What particles must be emitted by a nucleus for these transformations to occur?
    proton → neutron:
    neutron → proton:

Got It!

1. Write an equation describing the radioactive decay of each of the following isotopes. The type of decay is given in parentheses.
    $^{18}_{9}F$ (positron emission)

    $^{26}_{13}Al$ (electron capture)

    $^{35}_{16}S$ (beta emission)

    $^{218}_{84}Po$ (alpha emission)

    $^{119}_{50}Sn^{*}$ (gamma emission)

20-2. Rates of Nuclear Decay

What Do You Remember?

Review the chapter in your textbook on kinetics and the rates of reactions, and complete Table I. You should know how the integrated rate law combined with a graph can be used to identify the order of a reaction.

Table I. Rate Laws for Reactions

Reaction Order	Rate Law	Integrated Rate Law
0	$d[N]/dt =$	
1	$d[N]/dt =$	
2	$d[N]/dt =$	

Information

Radioactive iodine-131 commonly is used as part of the treatment for thyroid cancer because thyroid cells are unique in absorbing and concentrating iodine from the bloodstream. Thyroid cells need iodine for the synthesis of the thyroid hormones thyroxine and tri-iodothyronine.

Fig. 1   thyroxine   tri-iodothyronine

Iodine-131 decays by beta emission, and a typical treatment for thyroid cancer initially produces about 400×10^6 decay events per second. The level of radiation decreases with time as shown by the data in Figure 2 and Table II.

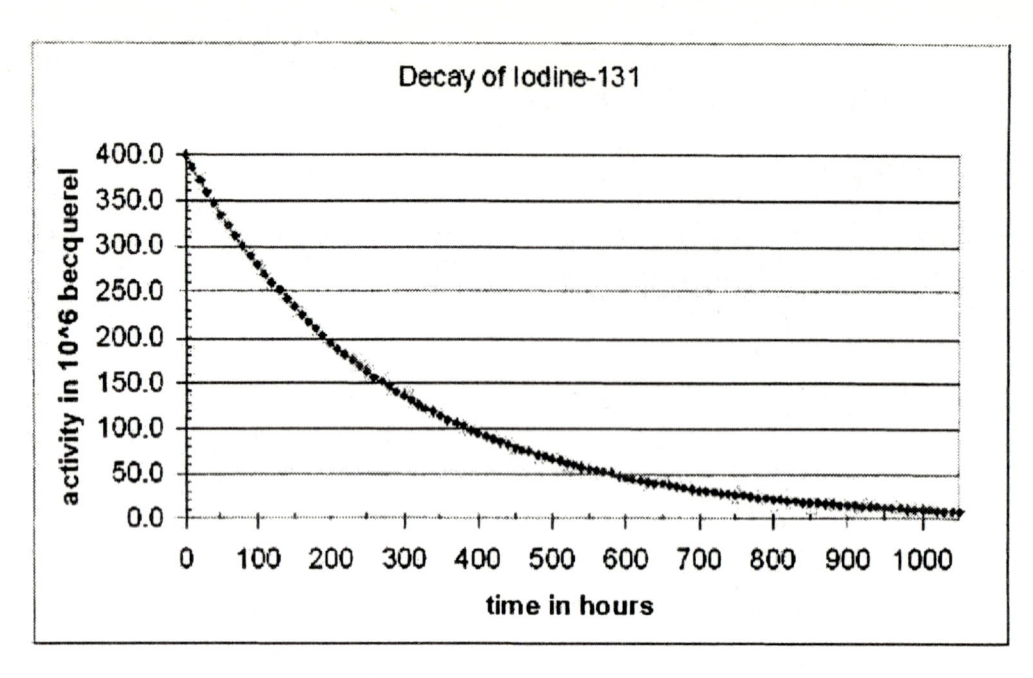

Fig. 2. Decay of Iodine-131.

Radioactivity is measured in units of decay events per second. The SI unit is the Becquerel (Bq), and 1 Bq = 1 s^{-1}. The becquerel is derived from Henri Becquerel, who shared the Nobel prize with Pierre and Marie Curie, for discovering radioactivity. Another commonly used unit for radioactivity is the curie (Ci), and 1 Ci = 3.7 x 10^8 s^{-1}, which is the number of decay events produced by 1 gram of radium-226.

Table II. Decay of Iodine-131

Time (hours)	Radioactivity (10^6 Bq)	Time (hours)	Radioactivity (10^6 Bq)
0	400	550	56
50	334	600	46
100	279	650	39
150	233	700	32
200	195	750	27
250	163	800	23
300	136	850	19
350	114	900	16
400	95	950	13
450	80	1000	11
500	66	1050	9

Exploration

1. Write the nuclear reaction equation for the decay of iodine-131 by beta particle emission.

2. Use the information in Figure 2 and Table II to determine whether the decay of iodine-131 is a zero, first, or second order reaction. State your conclusion and explain how it follows from the data. The information that you entered in Table I may be helpful.

3. Use the information in Figure 2 and Table II to determine
   (a) the rate constant for the decay, and

   (b) the half life of iodine-131 in days.

4. Explain how your responses to Exploration question (3) was obtained from the data.

Got It!

1. As a doctor planning a treatment you need to determine how many grams of iodine-131 must be absorbed by the thyroid to produce an initial activity of 400×10^6 Bq. The rate law that you entered in Table I will be helpful to you. How many grams are needed?

2. As a doctor you need to advise a patient on how long to refrain from certain activities until the radiation level subsides. How many days does it take for the radioactivity to fall to 4×10^6 Bq from the initial level of 400×10^6 Bq?

3. If the dose were doubled to produce an initial activity of 800×10^6 Bq, how many days would it take for the activity to fall to 4×10^6 Bq?

4. A patient had an initial dose of iodine-131 that produced an activity in the thyroid of 30 mCi (millicuries). At the most recent checkup, the activity was determined to be 1 mCi. How long ago was the initial dose administered?

21-1. Chemistry of the Main Group Elements: Why are some elements more special than others?

The Main Group Elements are those elements in Groups IA through 8A of the Periodic Table (columns 1, 2 and 13 through 18). Use your textbook and the internet as resources to explore the properties and applications of these elements. *Google* and *Wikipedia* should prove especially helpful.

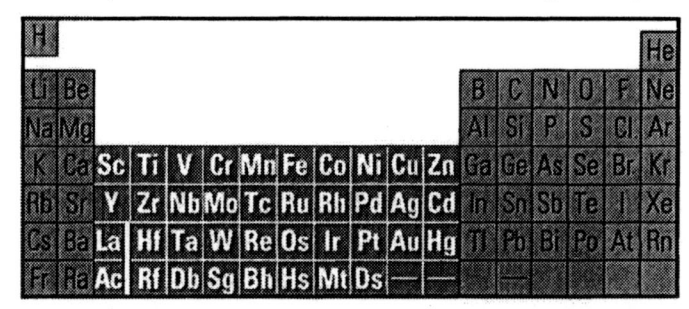

What do you know already?

Identify 5 main group elements that you are familiar with, and identify 2 uses for each of them.

Exploration

1. How many Main Group elements are there?

2. Which 5 elements are the most abundant on Earth?

3. What are 3 methods of extracting pure elements from their ores? Provide an example of each method.

4. Identify a use for one element in each group (IA through 8A). Do not pick one of the elements that you described in the *What do you know already?* section.

5. Which 5 elements do you see as the most essential for life? Explain.

6. Which 5 elements do you see as the most useful in technology and industry? Explain.

21-2. Electronic Structure and Properties

The properties of the Main Group elements depend on their electronic structures or electron configurations. In this activity you reflect on what you have learned previously about the periodic properties of the elements, and you identify the connection between the electron configuration of an element to its properties. The key ideas are the shell structure of the atom, the attraction of an electron by the positive nucleus, and the repulsion of an electron by the other electrons.

What do you remember?

Describe why atomic radii decrease across a Period and increase down a Group in the Periodic Table.

Exploration

1. Which Main Group elements are metals, and which are nonmetals?

2. Compared to metals, do nonmetals have high or low ionization potentials? Explain why.

3. Compared to metals, do nonmetals have high or low electron affinities? Explain why?

4. Compared to metals, do nonmetals have high or low electronegativities? Explain why?

5. For one element in each group (IA through 8A), identify a characteristic property of that element and explain the origin of that property in terms of its electron configuration, ionization potential, electron affinity, and electronegativity.

22-1 Transition Metals and Coordination Compounds: What's so special about transition metals?

The block of elements in the Periodic Table (columns 3 – 12) that have their origin in electrons filling the d-orbitals are called the *transition metals* or *transition elements*. Transition metals have many diverse and important properties. Some are very reactive and catalyze chemical reactions; others are inert and are used for coins and jewelry.

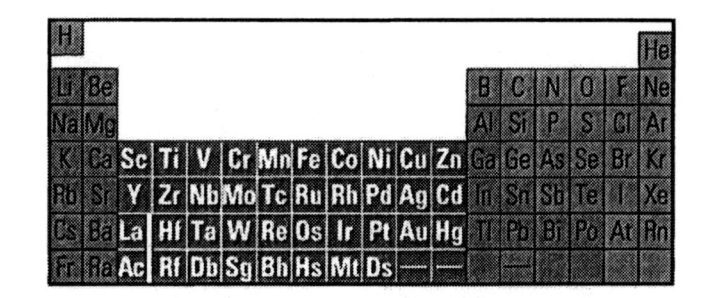

What do you know already?

Identify 5 transition metals that you are familiar with, and identify 2 uses for each of them.

Information

Many of the chemical properties of transition metals are related to their ability to form *coordinate covalent bonds* with molecules or ions that have lone pair electrons. Transition metals use d-orbitals to accept these electrons and form a chemical bond. Consequently, the bonding structure can involve more than the 8 electrons characteristic of the octet rule.

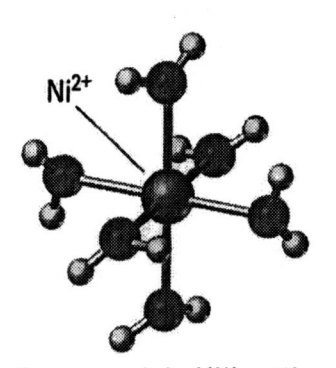

Fig. 1 Hexaquanickel(II) cation

Figure 1 shows the *complex ion* hexaaquanickel(II). It formed by 6 water molecules forming coordinate covalent bonds with a nickel +2 ion, which is the cation in the *coordination compound* hexaaquanickel(II) chloride, $[Ni(H_2O)_6]Cl_2$. The molecules or ions bonded to the central metal ion are called *ligands*.

Other coordination compounds and their components are listed in Table I, where CN = coordination number, Ox = oxidation number of the metal, Charge = total charge on the ligands plus metal.

Table I. Examples of Coordination Compounds

Coordination Compound	Complex Ion	Ligand	Ligand Geometry	Counter Ion	CN	Charge	Ox
$[Ni(H_2O)_6]Cl_2$	$[Ni(H_2O)_6]^{2+}$	H_2O	Octahedral	Cl^-	6	+2	+2
$K_3[Fe(CN)_6]$	$[Fe(CN)_6]^{3-}$					-3	+3
$[Co(NH_3)_3(OH)_3]$	None						
$Na_2[HgI_4]$	$[HgI_4]^{2-}$		Tetrahedral		4		

Exploration

1. What components of a coordination compound are present in all such compounds?

2. What additional component may be present in some coordination compounds?

3. Complete the missing entries in Table I. Note that the total charge on the ligands and metal can be identified from total charge on the counter ions.

4. What purpose is served by the square bracket in the chemical formula for a coordination compound?

5. When a coordination compound dissolves in water, it dissociates into the complex ion and the counter ions. What are the dissociation products for each coordination compound in Table I?

Got It!

1. Based on what you have seen in this activity, write definitions of the following terms.

    coordinate covalent bond

    coordination compound

    complex ion

    ligand

    coordination number

    counter ion

2. Describe what information in Table I can be used to determine the charge on the complex ion and the oxidation number of the metal.

Further Exploration

Use your textbook and the internet as resources to explore the properties and applications of transition metals and answer the following questions.

1. Identify 5 transition metals that are active in biological systems and describe the function of each.

2. Identify 5 transition metals or coordination compounds that are useful in technology and industry and describe the function of each.

22-2. Magnetism and Color in Coordination Compounds: Where does all that attraction and beauty come from?

Brilliant colors and magnetism are characteristic of many coordination compounds. Ligand-field theory (aka crystal-field theory) is a simple model that explains these properties. In ligand-field theory, the bonding between the ligands and the metal ion is considered to be electrostatic. The negative electrons on the ligands are attracted to the positive metal ion. The presence of the ligands alters the energy of the metal's d-orbitals. Unpaired electrons in the d-orbitals produce magnetism, and electronic transitions between the d-orbitals absorb light and cause the bright colors.

What do you remember?

Use the Aufbau Principle, Pauli Exclusion Principle, and Hunds Rule to predict which transition metals in Period 4 (Sc through Zn) have unpaired electrons.

Exploration – 1

According to the ligand-field model, the energies of the d-orbitals is a function of the relative positions of the ligands and the d-orbitals as shown in Figure 1.

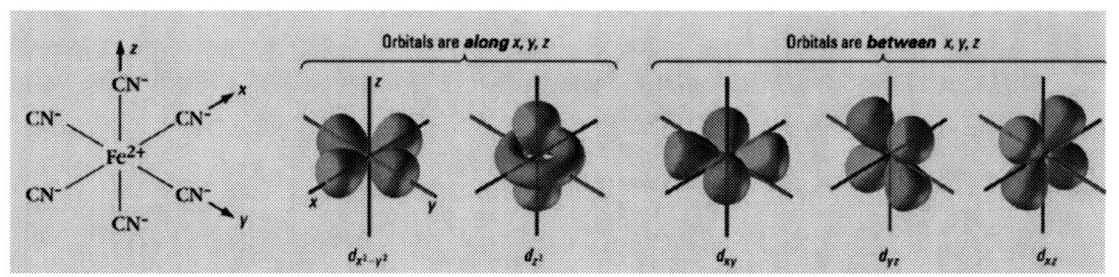

Fig. 1. The spatial arrangement of d-orbitals of an Fe2+ ion and the octahedral configuration of 6 CN- ligands. The ligands are aligned along the x, y, and z axes and interact most strongly with electrons in the 2 orbitals that also are aligned along the axes and not so strongly with electrons in the 3 orbitals that are aligned between the axes.

1.1. Which d-orbitals lie along the x, y, or z axes, and which lie between any two axes?

1.2. Which orbitals experience a direct head-on interaction with the ligands that also lie along the axes?

1.3. Which orbitals do not experience a direct head-on interaction with the ligands?

1.4. Based on the nature of the interactions that you identified above in Exploration questions (1.1) – (1.3), divide the five d-orbitals into two sets, and identify the set that has the stronger interactions with the ligands.

   Set 1:

   Set 2:

1.5. In a coordination compound with octahedral geometry, the energies of the two sets of orbitals depend upon the strength of the interactions with the ligands. The set experiencing the stronger interactions will be higher in energy because the electron-electron interactions are repulsive. Draw an energy-level diagram below to show the relative energies of the two sets of orbitals that you identified in Exploration question (4). Represent each of the 5 orbitals by a line along with the label for that orbital.

Exploration – 2

The diagram that you drew in response to Exploration question (1.5) is called a *ligand-field-splitting energy-level diagram*. The energy difference between the two sets of orbitals is referred to as Δo, where *o* refers to the octahedral geometry. Δo is called the *ligand-field* or *crystal-field* splitting, even though it has nothing to do with a crystal.

2.1. Which ligand shown in Figure 2 produces the larger ligand-field splitting?

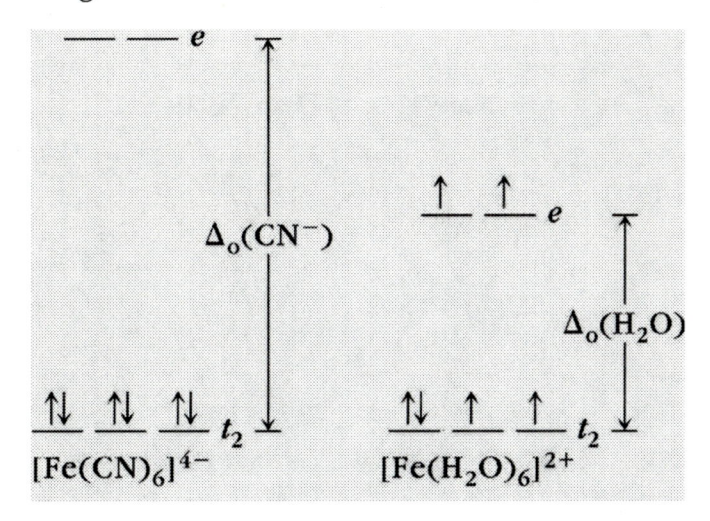

Fig. 2. Ligand- field splitting for cyanide and water. Cyanide is a strong field ligand and water is a weak field ligand

2.2. What is the difference between the ways electrons fill d-orbitals in the weak-field complex ion and in the strong-field complex ion shown in Figure 2?

2.3. Since electrons fill orbitals to produce the lowest energy state, why do you think electrons are unpaired in the free ion and in weak-field complex ion but pair-up in the strong field complex ion? Use the relative magnitudes of the ligand-field splitting Δo and the electron-electron repulsive energy in your answer.

2.4. Why do you think weak-field complex ions sometimes are called high spin, and strong-field complex ions sometimes are called low spin?

2.5. A spinning charge acts like an electromagnet with the polarity determined by the direction of spin. Why do you think the weak-field complex ion shown in Figure 2 is paramagnetic and the strong field complex ion is diamagnetic?

Exploration – 3

Light that is absorbed by an octahedral complex can be used to promote an electron from a t_2-orbital to an e-orbital. To do so, the energy in the photon, $h\nu$ or hc/λ, must equal the ligand-field splitting, Δo. Table I gives the approximate wavelength of light that is absorbed by 4 different cobalt complex ions.

Table I. Colors of solutions of Co^{3+} complexes.

Complex	Observed	Absorbed	Wavelength (nm)	Δo (J)
$[Co(NH_3)_6]^{3+}$	Yellow	Bluish violet	430	
$[Co(NH_3)_5NCS]^{2+}$	Orange	Bluish green	470	
$[Co(NH_3)_5H_2O]^{3+}$	Red	Green	500	
$[Co(NH_3)_5Cl]^{2+}$	Reddish purple	Yellowish green	522	

3.1. Why is the color of a solution different from the color of the light absorbed by it, as indicated in Table I?

3.2. Using the relationship between wavelength and photon energy, arrange the cobalt complexes in order of increasing energy of the photons absorbed.

3.3. Determine the value for the ligand field splitting for each of the cobalt complex ions in Table I from the energy of the photon absorbed, and add this information to the last column in Table I.

3.4. Draw d-orbital energy level diagrams for the 4 cobalt complex ions in Table I to show how the ligand-field splitting changes.

Got It!

Fe^{3+} forms octahedral complexes with NCS$^-$ and NO$_2^-$. One displays greater paramagnetism than the other.

1. Write the chemical formula for each of these complex ions.

2. Use the spectrochemical series to predict whether each complex ion is high spin or low spin.

3. Identify which complex ion is more paramagnetic. Explain.

4. Draw the ligand-field splitting diagram showing which orbitals are occupied by electrons for each complex ion.

5. Predict which complex ion will absorb light with the longer wavelength. Explain.